Ben Stacy Jerrik (Ed.)

Naviglio Pavese

Ben Stacy Jerrik (Ed.)

Naviglio Pavese

Navigli, Lombardy, Venice, Ticino (river), Regions of Italy, Lake Maggiore

Part Press

Imprint

Publisher:
Part Press is a trademark of
International Book Market Service Ltd., 17 Rue Meldrum, Beau Bassin, 1713-01 Mauritius
Email: info@bookmarketservice.com
Website: www.bookmarketservice.com

Published in 2012

Printed in: U.S.A., U.K., Germany. This book was not produced in Mauritius.

ISBN: 978-613-8-89284-7

Contents

Articles

References

Naviglio_Pavese

The **Naviglio Pavese** is one of the canals making up the Navigli system in Lombardy, Italy. Once navigable, it is 33 km long and connected the city of Milan to Pavia, and through a flight of six locks to the River Ticino.

Construction started in 1564, but was interrupted 20 years later just outside Milan due to technical problems: the lock there is still called *Conca Fallata*, which in Italian means "Failed Lock". Building resumed at the beginning of the 19th century and was completed in 1819.

The canal was finally closed to navigation in the 1960s, but work has recently started to restore it back to full navigation, a link in the project to connect Switzerland to Venice by inland waterway.

Canal

Canals are man-made channels for water. There are two types of canal:

1. Waterways: navigable transportation canals used for carrying ships and boats shipping goods and conveying people, further subdivided into two kinds:
    1. Those connected to existing lakes, rivers, or oceans. Included are inter-basin canals, such as the Suez Canal, Erie Canal, and the Panama Canal.
    2. Those connected in a city network: such as the *Canal Grande* and others of Venice Italy; the *gracht* of Amsterdam, and the waterways of Bangkok.
2. Aqueducts: water supply canals that are used for the conveyance and delivery of potable water for human consumption, municipal uses, and agriculture irrigation. Rills and acequias are small versions.

The Alter Strom, in the sea resort of Warnemünde, Germany.

The Royal Canal in Ireland.

Types of artificial waterways

The Danube-Black Sea Canal in Romania

Canals are created in one of three ways, or a combination of the three, depending on available water and available path:

- A canal can be created where no stream presently exists. Either the body of the canal is dug or the sides of the canal are created by piling dirt, stone, concrete, or other building materials. The water for the canal must be provided from an external source like other streams or reservoirs. Examples include canals that connect valleys over a higher body of land, like Canal du Midi and Canal de Briare.
- A stream can be *canalized* to make its navigable path more predictable and easier to maneuver. Canalization modifies the stream to more safely carry traffic by controlling the flow of the stream with dredging, damming, and modifying its path. Examples include Basse Saône, Canal de Mines de Fer de la Moselle, and Aisne River. Riparian zone restoration may be required.
- When a stream is too difficult to modify with *canalization*, a second stream can be created next to the existing stream. This is called a *lateral canal*. The existing stream usually acts as the water source and its banks provide a path for the new body. Examples include the Chesapeake and Ohio Canal, Canal latéral à la Loire, Garonne Lateral Canal, and Canal latéral à l'Aisne.

Smaller transportation canals can carry barges or narrowboats, while ship canals allow seagoing ships to travel to an inland port (e.g., Manchester Ship Canal), or from one sea or ocean to another (e.g., Caledonian Canal, Panama Canal).

Features

The flight of 16 consecutive locks at Caen Hill on the Kennet and Avon Canal, Wiltshire, England

At their simplest, canals consist of a trench filled with water. Depending on the stratum the canal passes through, it may be necessary to line the cut with some form of watertight material such as clay or concrete. When this is done with clay it is known as puddling.

Canals need to be level, and, while small irregularities in the lie of the land can be dealt with through cuttings and embankments, for larger deviations, other approaches have been adopted. The most common is the pound lock, which consists of a chamber within which the water level can be raised or lowered connecting either two pieces of canal at a different level or the canal with a river or the sea. When there is a hill to be climbed, flights of many locks in short succession may be used.

Prior to the development of the pound lock in 984AD in China by Chhaio Wei-Yo[1] and later in Europe in the 15th century, either flash locks consisting of a single gate were used or ramps, sometimes equipped with rollers, were used to change level. Flash locks were only practical where there was plenty of water available.

Locks use a lot of water, so builders have adopted other approaches. These include boat lifts, such as the Falkirk wheel, which use a caisson of water in which boats float while being moved between two levels; and inclined planes where a caisson is hauled up a steep railway.

A canal boat traverses the longest and highest aqueduct in the UK, at Pontcysyllte in Denbighshire, Wales

To cross a stream or road, the solution is usually to bridge with an aqueduct. To cross a wide valley (where the journey delay caused by a flight of locks at either side would be unacceptable) the centre of the valley can be spanned by an aqueduct - a famous example in Wales is the Pontcysyllte aqueduct across the valley of the River Dee.

Another option for dealing with hills is to tunnel through them. An example of this approach is the Harecastle Tunnel on the Trent and Mersey Canal. Tunnels are only practical for smaller canals.

Some canals attempted to keep changes in level down to a minimum. These canals known as contour canals would take longer winding routes, along which the land was a uniform altitude. Other generally later canals took more direct routes requiring the use of various methods to deal with the change in level.

Canals have various features to tackle the problem of water supply. In some cases such as the Suez Canal the canal is simply open to the sea. Where the canal is not at sea level a number of approaches have been adopted. Taking water from existing rivers or springs was an option in some cases, sometimes supplemented by other methods to deal with seasonal variations in flow. Where such sources were unavailable, reservoirs - either separate from the canal or built into its course - and back pumping were used to provide the required water. In other cases, water pumped from mines was used to feed the canal. In certain cases, extensive "feeder canals" were built to bring water from sources located far from the canal.

Where large amounts of goods are loaded or unloaded such as at the end of a canal a canal basin may be built. This would normally be a section of water wider than the general canal. In some cases, the canal basins contain wharfs and cranes to assist with movement of goods.

When a section of the canal needs to be sealed off so it can be drained for maintenance stop planks are frequently used. These consist of planks of wood placed across the canal to form a dam. They are generally placed in pre existing grooves in the canal bank. On more modern canals, "guard locks" or gates were sometimes placed to allow a section of canal to be quickly closed off, either for maintenance, or to prevent a major loss of water due to a canal breach.

History

Ancient canals

The Grand Canal of China at Suzhou

The oldest known canals were irrigation canals, built in Mesopotamia circa 4000 BC, in what is now modern day Iraq and Syria. The Indus Valley Civilization, Ancient India, (circa 2600 BC) had sophisticated irrigation and storage systems developed, including the reservoirs built at Girnar in 3000 BC.[2] In Egypt, canals date back at least to the time of Pepi I Meryre (reigned 2332–2283 BC), who ordered a canal built to bypass the cataract on the Nile near Aswan.[3]

In ancient China, large canals for river transport were established as far back as the Warring States (481–221 BC), the longest one of that period being the Hong Gou (Canal of the Wild Geese), which according to the ancient historian Sima Qian connected the old states of Song, Zhang, Chen, Cai, Cao, and Wei.[4] By far the longest canal was the Grand Canal of China, still the longest canal in the world today, and the oldest extant one. It is 1794 kilometres (**unknown operator: u'strong'** mi) long and was built to carry the Emperor Yang Guang between Beijing and Hangzhou. The project began in 605 and was completed in 609, although much of the work combined older canals, the oldest section of the canal existing since at least 486 BC. Even in its narrowest urban sections it is rarely less than 30 metres (**unknown operator: u'strong'** ft) wide.

Greek engineers were the first to use canal locks, by which they regulated the water flow in the Ancient Suez Canal as early as the 3rd century BC.[5] [6] [7]

Middle Age

In the Middle Ages, water transport was cheaper and faster than transport overland. This was because roads were unpaved and in poor condition and greater amounts could be transported by ship. The first artificial canal in Christian Europe was the Fossa Carolina built at the end of the 8th century under personal supervision of Charlemagne.

More lasting and of more economic impact were canals like the Naviglio Grande built between 1127 and 1257 to connect Milan with the Ticino River. The Naviglio Grande is the most important of the lombard "navigli"[8] and the oldest functioning canal in Europe.

Later, canals were built in the Netherlands and Flanders to drain the polders and assist the transportation of goods.

Thal Canal, Punjab, Pakistan.

Canal building was revived in this age because of commercial expansion from the 12th century. River navigations were improved progressively by the use of single, or flash locks. Taking boats through these used large amounts of water leading to conflicts with watermill owners and to correct this, the pound or chamber lock first appeared, in 10th century in China and in Europe in 1373 in Vreeswijk, Netherlands.[9] Another important development was the mitre gate, which was, it is presumed, introduced in Italy by Bertola da Novate in the 16th century. This allowed wider gates and also removed the height restriction of guillotine locks.

To break out of the limitations caused by river valleys, the first summit level canals were developed with the Grand Canal of China in 581–617 AD whilst in Europe the first, also using single locks, was the Stecknitz Canal in Germany in 1398.

Early Modern Age

The first to use pound locks was the Briare Canal connecting the Loire and Seine (1642), followed by the more ambitious Canal du Midi (1683) connecting the Atlantic to the Mediterranean. This included a staircase of 8 locks at Béziers, a 157 metres (**unknown operator: u'strong'** ft) tunnel and three major aqueducts.

Canal building progressed steadily in Germany in the 17th and 18th centuries with three great rivers, the Elbe, Oder and Weser being linked by canals. In post-Roman Britain, the first canal built appears to have been the Exeter Canal, which opened in 1563. The oldest canal built for industrial purposes in North America is Mother Brook in Dedham, MA. It was constructed in 1639 to provide water power for mills.

In Russia, the Volga-Baltic Waterway, a nationwide canal system connecting the Baltic and Caspian seas via the Neva and Volga rivers, was opened in 1718.

Industrial Revolution

- See also: History of the British canal system
- See also: History of turnpikes and canals in the United States

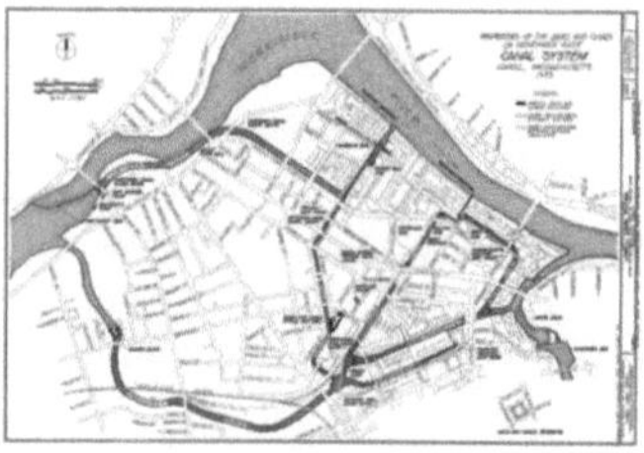

Lowell's power canal system

Canals were important for industrial development. The greatest stimulus to canal system building came from the Industrial Revolution with its need for cheap transport of unprecedented quantities of raw materials and manufactured items.

In Europe, particularly Britain and Ireland, and then in the young United States and the Canadian colonies, inland canals preceded the development of railroads (1780-1840) during the earliest phase of the Industrial Revolution. The opening of the Sankey Canal in 1757, followed by the Bridgewater Canal in 1761, which halved the price of coal in Liverpool and Manchester England, respectively, triggered a period of "canal mania" in Britain so that between 1760 and 1820 over one hundred canals were built.

As well as industrial uses, the Royal Military Canal on the Romney Marsh was built so as create a barrier against invading troops, and hiding places for British troops during the Napoleonic Wars.[10]

Aqueduct over the Mohawk River at Rexford, New York, one of 32 aqueducts on the Erie Canal.

Canal companies were initially chartered by individual states in the United States. These early canals were constructed, owned, and operated by private joint-stock companies. Three were completed when the War of 1812 broke out; these were the Santee Canal (opened 1800) in South Carolina, the Middlesex Canal (opened 1802) in Massachusetts and the Dismal Swamp Canal (opened 1805) in Virginia. The Erie Canal (opened 1825) was chartered and owned by the state of New York and financed by bonds bought by private investors. The Erie canl runs about 363 miles (**unknown operator: u'strong'** km) from Albany, New York, on the Hudson River to Buffalo, New York, at Lake Erie. The Hudson River connects Albany to the Atlantic port of New York City and the Erie Canal completed a navigable water route from the Atlantic Ocean to the Great Lakes. The canal contains 36 locks and encompasses a total elevation differential of around 565 ft. (169 m). The Erie Canal with its easy connections to most of the U.S. mid-west and New York City soon quickly paid back all its invested capital (USD $7 million) and started turning a profit. By cutting transportation costs in half or more it became a large profit center for Albany, New York and New York City as it allowed the cheap transportation of many of the agricultural products grown in the mid west of the United States to the rest of the world. From New York City these agricultural products could easily be shipped to other U.S. states or to Europe, etc. Assured of a market for their farm products the settlement of the U.S. mid-west was greatly accelerated by the Erie Canal. The profits generated by the Erie Canal project started a canal building boom in the United States that lasted till about 1850 when railroads started becoming seriously competitive in price and convience. The Blackstone Canal (finished in 1828) in Massachusetts and Rhode Island fulfilled a similar role in the early industrial revolution between 1828–48. The Blackstone Valley was a major contributor of the American Industrial Revolution where Samuel Slater built his first textile mill.

In addition to their transportation purposes, parts of the United States, particularly in the Northeast, had enough fast-flowing rivers that water power was the primary means of powering factories (usually textile mills) until after the American Civil War. For example, Lowell, Massachusetts, considered to be "The Cradle of the American Industrial Revolution," has 6 miles (**unknown operator: u'strong'** km) of canals, built from around 1790 to 1850, that provided water power and a means of transportation for the city. The output of the system is estimated at 10,000 horsepower.[11] Other cities with extensive power canal systems include Lawrence, Massachusetts, Holyoke, Massachusetts, Manchester, New Hampshire, and Augusta, Georgia.

19th century

Competition from the railway network from the 1830s, and in the 20th century the roads, made the smaller canals obsolete for most commercial transportation, and many of the British canals fell into decay. Only the Manchester Ship Canal and the Aire and Calder Canal bucked this trend. Yet in other countries canals grew in size as construction techniques improved. During the 19th century in the US, the length of canals grew from 100 miles (**unknown operator: u'strong'** km) to over 4,000, with a complex network making the Great Lakes navigable, in conjunction with Canada, although some canals were later drained and used as railroad rights-of-way.

US canals circa 1825

In the United States, navigable canals reached into isolated areas and brought them in touch with the world beyond. By 1825 the Erie Canal, 363 miles (**unknown operator: u'strong'** km) long with 82 locks, opened up a connection from the populated Northeast to the Great Lakes. Settlers flooded into regions serviced by such canals, since access to markets was available. The Erie Canal (as well as other canals) was instrumental in lowering the differences in commodity prices between these various markets across America. The canals caused price convergence between different regions because of their reduction in transportation costs, which allowed Americans to ship and buy goods from farther distances for much lower prices compared to before. Ohio built many miles of canal, Indiana had working canals for a few decades, and the Illinois and Michigan Canal connected the Great Lakes to the Mississippi River system until replaced by a channelized river waterway.

Three major canals with very different purposes were built in what is now Canada. The first Welland Canal, which opened in 1829 between Lake Ontario and Lake Erie, bypassing Niagara Falls and the Lachine Canal (1825), which allowed ships to skirt the nearly impassable rapids on the St. Lawrence River at Montreal were built for commerce. The Rideau Canal, completed in 1832, connects Ottawa, on the Ottawa River to Kingston, Ontario on Lake Ontario. The Rideau Canal was built as a result of the War of 1812 to provide military transportation between the British colonies of Upper Canada and Lower Canada as an alternative to part of the St. Lawrence River, which was susceptible to blockade by the United States.

In France, a steady linking of all the river systems—Rhine, Rhône, Saône, and Seine—and the North Sea was boosted in 1879 by the establishment of the Freycinet gauge, which specified the minimum size of locks so that canal traffic doubled in the first decades of the 20th century.[12]

Many notable sea canals were completed in this period, starting with the Suez Canal (1869), and the Kiel Canal (1897), which carries tonnage many times that of most other canals, though the Panama Canal was not opened until 1914.

In the 19th century, a number of canals were built in Japan including the Biwako canal and the Tone canal. These canals were partially built with the help of engineers from the Netherlands and other countries.[13]

Modern uses

Canals can disrupt water circulation in marsh systems.

Large-scale ship canals such as the Panama Canal and Suez Canal continue to operate for cargo transportation, as do European barge canals. Due to globalization, they are becoming increasingly important, resulting in expansion projects such as the Panama Canal expansion project.

The narrow early industrial canals, however, have ceased to carry significant amounts of trade and many have been abandoned to navigation, but may still be used as a system for transportation of untreated water. In some cases railways have been built along the canal route, an example being the Croydon Canal.

A movement that began in Britain and France to use the early industrial canals for pleasure boats, such as hotel barges, has spurred rehabilitation of stretches of historic canals. In some cases, abandoned canals such as the Kennet and Avon Canal have been restored and are now used by pleasure boaters. In Britain, canalside housing has also proven popular in recent years.

The Seine-Nord Europe Canal is being developed into a major transportation waterway, linking France with Belgium, Germany, and the Netherlands.

Canals have found another use in the 21st century, as easements for the installation of fibre optic telecommunications network cabling, avoiding having them buried in roadways while facilitating access and reducing the hazard of being damaged from digging equipment.

Canals are still used to provide water for agriculture. An extensive canal system exists within the Imperial Valley in the Southern California desert to provide irrigation to agriculture within the area.

Cities on water

An intersection of two canals (Grachten) in Amsterdam, Netherlands.

Canals are so deeply identified with Venice that many canal cities have been nicknamed "the Venice of...". The city is built on marshy islands, with wooden piles supporting the buildings, so that the land is man-made rather than the waterways. The islands have a long history of settlement; by the 12th century, Venice was a powerful city state.

Amsterdam was built in a similar way, with buildings on wooden piles. It became a city around 1300.

Other cities with extensive canal networks include: Alkmaar, Amersfoort, Bolsward, Brielle, Delft, Den Bosch, Dokkum, Dordrecht, Enkhuizen, Franeker, Gouda, Haarlem, Harlingen, Leeuwarden, Leiden, Sneek and Utrecht in the Netherlands, Brugge and Gent in Flanders, Belgium, Birmingham in England — which has 35 miles of canals, to Venice's 26 miles —, Saint Petersburg in Russia, Hamburg in Germany, Berlin in Germany, Fort Lauderdale, Florida, and Cape Coral, Florida in the United States.

Liverpool Maritime Mercantile City is a UNESCO World Heritage Site near the centre of Liverpool, England, where a system of intertwining waterways and docks is now being developed for mainly residential and leisure use.

Griboyedov Canal in St. Petersburg, Russia.

Canal Estates (commonly known as bayous) are a form of subdivision popular in cities like Miami, Florida, Texas City, Texas and the Gold Coast, Queensland; the Gold Coast has over 700 km of residential canals. Wetlands are difficult areas upon which to build housing estates, so dredging part of the wetland down to a navigable channel provides fill to build up another part of the wetland above the flood level for houses. Land is built up in a finger pattern that provides a suburban street layout of waterfront housing blocks.

Boats

Two Panamax ships in the Miraflores Locks on the Panama Canal, Panama.

Inland canals have often had boats specifically built for them. An example of this is the British narrowboat, which is up to 72 feet (**unknown operator: u'strong'** m) long and 7 feet (**unknown operator: u'strong'** m) wide and was primarily built for British Midland canals. In this case the limiting factor was the size of the locks. This is also the limiting factor on the Panama canal where Panamax ships are limited to a length of 294.1 m (**unknown operator: u'strong'** ft) and a width of 32.3 m (**unknown operator: u'strong'** ft). For the lockless Suez Canal the limiting factor for Suezmax ships is generally draft, which is limited to 16 m (**unknown operator: u'strong'** ft). At the other end of the scale, tub-boat canals such as the Bude Canal were limited to boats of under 10 tons for much of their length due to the capacity of their inclined planes or boat lifts. Most canals have a limit on height imposed either by bridges or by tunnels.

Lists of canals

- **Europe**
  - Canals of France
  - Canals of Amsterdam
  - Canals of Germany
  - Canals of Ireland
  - Canals of Russia
  - Canals of the United Kingdom
- **North America**
  - Canals of Canada
  - Canals of the United States

Amsterdam *gracht*

See also

- Barges of all types
- British Waterways
- Canal tunnel
- Channel
- Ditch
- Environment Agency
- History of the British canal system
- Horse-drawn boat
- Infrastructure
- Irrigation district
- Lists of canals
- List of navigation authorities in the United Kingdom
- List of waterways
- List of waterway societies in the United Kingdom
- Lock
- Navigable aqueduct
- Navigation authority
- Narrowboat
- Ship canal
- Roman canals - (*Torksey*)
- Volumetric flow rate
- Water bridge
- Waterscape
- Water transportation
- Waterway
- Waterway restoration
- Waterways in the United Kingdom
- Weigh lock

References

Notes

[1] Hadfield 1986, p. 22.
[2] Rodda 2004, p. 161.
[3] Hadfield 1986, p. 16.
[4] Needham 1971, p. 269.
[5] Moore, Frank Gardner (1950): "Three Canal Projects, Roman and Byzantine", *American Journal of Archaeology*, Vol. 54, No. 2, pp. 97–111 (99–101)
[6] Froriep, Siegfried (1986): "Ein Wasserweg in Bithynien. Bemühungen der Römer, Byzantiner und Osmanen", *Antike Welt*, 2nd Special Edition, pp. 39–50 (46)
[7] Schörner, Hadwiga (2000): "Künstliche Schiffahrtskanäle in der Antike. Der sogenannte antike Suez-Kanal", *Skyllis*, Vol. 3, No. 1, pp. 28–43 (33–35)
[8] Calvert 1963, p. .
[9] (PDF) *The International Canal Monuments List* (http://www.icomos.org/studies/canals.pdf), , retrieved 2008-10-08
[10] "History" (http://www.royalmilitarycanal.com/pages/history.asp). The Royal Military Canal. 1909-12-15. . Retrieved 2011-08-05.
[11] *Lowell National Historical Park — Lowell History Prologue* (http://www.nps.gov/archive/lowe/loweweb/Lowell History/prologue.htm), , retrieved 2008-10-08
[12] Edwards 2002, p. .
[13] Hadfield 1986, p. 191.

Bibliography

- Calvert, Roger (1963), *Inland Waterways of Europe*, George Allen and Unwin
- Edwards-May, David (2008), *European Waterways - map and concise directory, 3rd edition*, Euromapping
- Hadfield, Charles (1986), *World Canals: Inland Navigation Past and Present*, David and Charles, ISBN 0-7153-8555-0
- Needham, J (1971), *Science and Civilisation in China*, C.U.P. Cambridge
- Rodda, J. C. (2004), *The Basis of Civilization - Water Science?*, International Association of Hydrological Sciences

External links

- British Waterways' leisure website - Britain's official guide to canals, rivers and lakes (http://www.waterscape.com/)
- Leeds Liverpool Canal Photographic Guide (http://www.towpathtreks.co.uk/)
- Triumphs of Canal Building (http://www.engwonders.orgfree.com/e043.html)
- Information and Boater's Guide to the New York State Canal System (http://www.nycanals.com/)
- "Canals and Navigable Rivers" by James S. Aber, Emporia State University (http://academic.emporia.edu/aberjame/wetland/canal/canals.htm)
- National Canal Museum (USA) (http://www.canals.org/)
- London Canal Museum (UK) (http://www.canalmuseum.org.uk/)
- Canals in Amsterdam (http://www.world-city-photos.org/Amsterdam/photos/Canals_and_Bridges/)
- Canal du Midi (http://www.canaldumidi.com)
- Canal des Deux Mers (http://www.canaldumidi.com/Canal-des-Deux-Mers.php)
- Canal flow measurement using a sensor (http://www.usbr.gov/pmts/hydraulics_lab/water/bubbler/index.htm).

Navigli

The **navigli** was a system of navigable and interconnected canals around Milan, Italy.

The system consisted of five canals: Naviglio Grande, Naviglio Pavese, Naviglio Martesana, Naviglio di Paderno, Naviglio di Bereguardo. The first three were connected through Milan via the *Fossa Interna*, also known as the Inner Ring. The urban section of the Naviglio Martesana was covered over at the beginning of the 1930s, together with the entire Inner Ring, thus sounding the death knell for the north-eastern canals. Commercial carrying continued on the Naviglio Grande, but the decline was steady and by the sixties it was over for good.

Today the canals are mostly derelict, unnavigable, or used for irrigation. However, plans are afoot to restore the Naviglio Grande and the Naviglio Pavese.

External links

- Martesana Canal [1] - Historic and present pictures
- Inner ring [2] - Historic and present pictures
- Darsena [3] - Historic and present pictures

References

[1] http://www.canoamartesana.it/galleria.asp?act=foto&idgalleria=471
[2] http://www.canoamartesana.it/galleria.asp?act=foto&idgalleria=474
[3] http://www.canoamartesana.it/galleria.asp?act=foto&idgalleria=476

Lombardy

Lombardy
Lombardia

— Region of Italy —

Flag

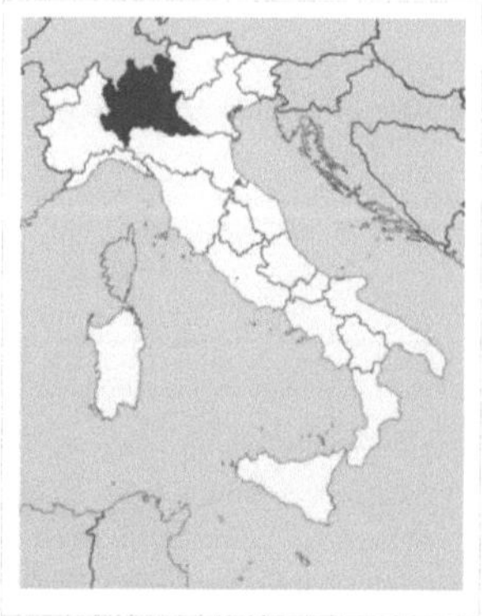

Country	Italy
Capital	Milan
Government	
• President	Roberto Formigoni (PDL)
Area	
• Total	23844 km^2 (**unknown operator: u'strong'** sq mi)
Population (25 March 2011)	
• Total	9939193
• Density	**unknown operator: u'strong'**/km^2 (**unknown operator: u'strong'**/sq mi)
Demonym	Lombard
Time zone	CET (UTC+1)
• Summer (DST)	CEST (UTC+2)
GDP/ Nominal	€ 328.2[1] billion (2008)
GDP per capita	€ 33,500[2] (2008)
NUTS Region	ITC
Website	www.regione.lombardia.it [3]

Lombardy (Italian: *Lombardia* Italian pronunciation: [lombar'diːa], Western Lombard: *Lumbardìa*, Eastern Lombard: *Lombardia*) is one of the 20 regions of Italy. The capital is Milan. One-sixth of Italy's population lives in Lombardy and about one fifth of Italy's GDP is produced in this region, making it the most populous and richest region in the country and one of the richest in the whole of Europe.[4] Major tourist destinations in the region include the historic, cultural and artistic cities of Milan (which is Italy's second top tourist destination),[5] Brescia, Mantua, Pavia, Como, Cremona, Bergamo, Sondrio, Lecco, Lodi, Varese, Monza, and the lakes of Garda, Como, Maggiore, and Iseo.

The official language, as in the rest of Italy, is Italian. The traditional local languages are the various dialects of Lombard (Western Lombard and Eastern Lombard), as well as some dialects of Emilian, spoken in some parts of the provinces of Mantua, Pavia, and Cremona. According to Istat, almost 27% of Lombards are bilingual with Lombard and Italian languages; 9.1% are monolingual in Lombard and 57.6% are monolingual in Italian.[6]

Geography

Satellite image of Lombardy

Lombardy is bordered by Switzerland (north: Canton Ticino and Canton Graubünden) and by the Italian regions of Emilia-Romagna (south), Trentino-Alto Adige/Südtirol and the Veneto (east), and Piedmont (west). Three distinct natural zones can be fairly easily distinguished in the Lombardy region: mountains, hills and plains – the latter being divided in *Alta* (high plains) and *Bassa* (low plains).

The most important mountainous area is an Alpine zone including the Lepontine and Rhaetian Alps, (Piz Bernina – La Spedla, 4,020 m), the Bergamo Alps, the Ortler Alps and the Adamello massif; it is followed by an Alpine foothills zone Prealpi, which include the main peaks are the Grigna Group (2,410 m), Resegone (1,875 m) and Presolana (2,521 m). The great Lombard lakes, all of glacial origin lie in this zone. From west to east these are Lake Maggiore, Lake Lugano (only a small part is Italian), Lake Como, Lake Iseo, Lake Idro, then Lake Garda, the largest in Italy. South of the Alps lie the hills characterized by a succession of low heights of morainic origin, formed during the last Ice Age and small barely fertile plateaux, with typical heaths and conifer woods. A minor mountainous area lies south of the Po, in the Appennines range.

Panorama view: *Bergamasque Alps*

The plains of Lombardy, formed from alluvial deposits, can be divided into the *Alta* – an upper, permeable ground zone in the north and a lower zone characterized – the *Bassa* – by the so-called line of *fontanili* (the spring waters rising on impermeable ground). Anomalous compared with the three distinctions already made is the small region of the Oltrepò Pavese, formed by the Apennine foothills beyond the Po River. A large number of rivers, all direct or indirect tributaries of the Po, cross the plains of Lombardy. Major rivers, flowing NW to SE, are the Ticino, the outlet of Lake Maggiore, the Olona, the Lambro, the Adda, outlet of Lake Como, the Mincio, outlet of Lake Garda, and the Oglio, the Lake Iseo outflow. There is a wide network of canals for irrigation purposes. In the plains, intensively cultivated for centuries, little of the original environment remains. The rare elm, alder, sycamore, poplar, willow and hornbeam woods and heaths are covered now by several protected areas. In the area of the great Alpine foothills lakes, however, grow olive trees, cypresses and larches, as well as varieties of subtropical flora such as magnolias, azaleas, acacias, etc. The mountains area is characterized by the typical vegetation of the whole range of the Italian Alps. At a lower levels (up to approximately 1,100 m) oak woods or broadleafed trees grow; on the mountain slopes (up to 2,000–2,200 m) beech trees grow at the lowest limits, with conifer woods higher up. Shrubs such as rhododendron, dwarf pine and juniper are native to the summital zone (beyond 2,200 m).

Panorama view: *Lake Maggiore*

The climate of this region is continental, though with variations depending on altitude or the presence of inland waters. The continental nature of the climate is more accentuated on the plains, with high annual temperature changes (at Milan an average January temperature is 1.5 °C (**unknown operator: u'strong'** °F) and 24 °C (**unknown operator: u'strong'** °F) in July), and thick fog between October and February. The Alpine foothills lakes exercise a mitigating influence, permitting the cultivation of typically Mediterranean produce (olives, citrus fruit). In the Alpine zone, the valley floor is relatively mild in contrast with the colder higher areas (Bormio, 1,225 m, −1.4 °C (**unknown operator: u'strong'** °F) average in January, 17.3 °C (**unknown operator: u'strong'** °F) in July). Precipitations are more frequent in the Prealpine zone (up to 1,500–2,000 mm annually) than on the plains and Alpine zones (600 mm to 850 mm annually). The numerous species of endemic flora (the Lombard native species), typical mainly of the Lake Como area, include some kinds of saxifrage, the Lombard garlic, groundsels bellflowers and the cottony bellflowers. Lombardy counts many protected areas: the most important are the Stelvio National Park (the largest Italian natural park), with typically alpine wildlife: red deer, roe-deer, ibex, chamois, foxes, ermine and also golden eagles; and the Ticino Valley Natural Park, instituted in 1974 on the Lombard side of the Ticino River to protect and conserve one of the last major examples of fluvial forest in Northern Italy.

History

Early history

Further information: Remedello culture, Polada culture, and Golasecca culture

Further information: Insubres, Cenomani, Raeti, and Camunni

Further information: Cisalpine Gaul

Roman ruins of the Imperial palace of Mediolanum

The area of current Lombardy was settled at least since the 2nd millennium BC, as shown by the archaeological findings of ceramics, arrows, axes and carved stones. In the following centuries it was inhabited by different peoples amongst whom the Etruscans, who founded the city of Mantua and spread the use of writing; later, starting from the 5th century BC, the area was invaded by Celtic – Gallic tribes. These people settled in several cities (including Milan) and extended their rule to the Adriatic Sea. Their development was halted by the Roman expansion in the Po Valley from the 3rd century BC onwards: after centuries of struggle, in 194 BC the entire area of what is now Lombardy became a Roman province with the name of Gallia Cisalpina ("Gaul on the nearer side of the Alps"). The Roman culture and language overwhelmed the former civilization in the following

Ancient Celtic artifacts retrieved in Lombardy[7]

years, and Lombardy became one of the most developed and rich areas of Italy with the construction of a wide array of roads and the development of agriculture and trade. Important figures like Pliny the Elder (in Como) and Virgil (in Mantua) were born here. In late antiquity the strategic role of Lombardy was emphasized by the temporary moving of the capital of the Western Empire to Mediolanum (Milan). Here, in 313 AD, emperor Constantine issued the famous edict that gave freedom of confession to all religions within the Empire.

Middle Ages

Further information: Gothic War (535–554), Battle of Taginae, and Lombards

Further information: Guelphs and Ghibellines

Further information: Duchy of Milan, Wars in Lombardy, and Venetian Republic

King Alboin led the Lombard migration into the Po Valley and made Pavia the capital

During and after the fall of the Western Empire, Lombardy suffered heavily from destruction brought about by a series of invasions by tribal peoples. The last and most effective was that of the Lombards, or Longobardi, who came around 570s and whose long-lasting reign (whose capital was set in Pavia) gave the current name to the region. There was a close relationship between the Frankish, Bavarian and Lombard nobility for many centuries. After the initial struggles, relationships between the Lombard people and the Latin-speaking people improved. In the end, the Lombard language and culture assimilated with the Latin culture, leaving evidence in many names, the legal code and laws among other things. The end of Lombard rule came in 774, when the Frankish king Charlemagne conquered Pavia and annexed the Kingdom of Italy (mostly northern and central Italy) to his empire. The former Lombard dukes and nobles were replaced by other German vassals, prince-bishops or marquises. The 11th century marked a significant boom in the region's economy, due to improved trading and, mostly, agricultural conditions. In a similar way to other areas of Italy, this led to a growing self-acknowledgement of the cities, whose increasing richness made them able to defy the traditional feudal supreme power, represented by the German emperors and their local legates. This process reached its apex in the 12th and 13th centuries, when different Lombard Leagues formed by allied cities of Lombardy, usually led by Milan, managed to defeat the Hohenstaufen Emperor Frederick I, at Legnano, and his grandson Frederick II, at Parma.

The Coronation ceremony of Gian Galeazzo Visconti in the Basilica of Sant'Ambrogio

This did not prevent other important Lombard centres, like Cremona (then rivalling Milan for size and wealth) and others, from supporting the imperial power if this could grant them an immediate advantage. Taking advantage of the flourishing agriculture, the area around the Po River, together with Venice and Tuscany, continued to expand its industry and commerce until it became the economic centre of the whole of Europe. The enterprising class of the communes extended its trade and banking activities well into northern Europe: "Lombard" designated the merchant or banker coming from northern Italy (see, for instance, Lombard Street in London). The name "Lombardy" came to designate the whole of Northern Italy until the 15th century and sometimes later. From the 14th century onwards, the instability created by the unceasing internal and external struggles ended in the creation of noble seignories, the most significant of which were those of the Viscontis (later Sforzas) in Milan and of the Gonzagas in Mantua. In the 15th century the Duchy of Milan was a major political, economical and military force at the European level. Milan and Mantua became two centres of the Renaissance whose culture, with men like Leonardo da Vinci and Mantegna, and pieces of art were highly regarded (for example, Leonardo da Vinci's *The Last Supper*). This richness, however, attracted the now more organized armies of national powers like France and Austria, which waged a lengthy battle for Lombardy in the late 15th-early 16th century.

Modern era

Further information: War of the Spanish succession and Kingdom of Italy (Napoleonic)

Further information: Five Days of Milan and Second Italian War of Independence

Further information: Italian Social Republic

Historical image of Villa Reale di Monza, official summer residence of the Austrian viceroys of Lombardy-Venetia, part of Austrian Empire

After the decisive Battle of Pavia, the Duchy of Milan became a possession of the Habsburgs of Spain: the new rulers did little to improve the economy of Lombardy, instead imposing a growing series of taxes needed to support their unending series of European wars. The eastern part of modern Lombardy, with cities like Bergamo and Brescia, was under the Republic of Venice, which had begun to extend its influence in the area from the 14th century onwards (*see also Italian Wars*). Pestilences (like that of 1628/1630,[8] described by Alessandro Manzoni in his *I Promessi Sposi*) and the generally declining conditions of Italy's economy in the 17th and 18th centuries halted the further development of Lombardy. In 1706 the Austrians came to power and introduced some economical and social measures which granted a certain recovery. Their rule was smashed in the late 18th century by the French armies, however, and with the formation of the Napoleonic Empire, Lombardy became one of the semi-independent province of Napoleonic France. The restoration of Austrian rule in 1815, in the form of the puppet state called Kingdom of Lombardy–Venetia, had however to contend with new social ideals introduced by the Napoleonic era.

Lombardy became one of the intellectual centres leading to Italian unification. The popular republic of 1848 was short-lived, its suppression leading to renewed Austrian rule. This came to a decisive end when Lombardy was annexed to the Kingdom of Italy 1859 as a result of the Second Italian Independence War. When annexed to the Kingdom of Italy in 1859 Lombardy achieved its present-day territorial shape by adding the Oltrepò Pavese (formerly the southern part of Novara's Province) to the province of Pavia. Starting from the late 19th century, and with a boom after World War II, Lombardy sharpened its status of richest and most industrialized region of Italy.

Demographics

Historical populations

Year	Pop.	±%
1861	3160000	—
1871	3529000	+11.7%
1881	3730000	+5.7%
1901	4314000	+15.7%
1911	4889000	+13.3%
1921	5186000	+6.1%
1931	5596000	+7.9%
1936	5836000	+4.3%
1951	6566000	+12.5%
1961	7406000	+12.8%
1971	8543000	+15.4%
1981	8892000	+4.1%
1991	8856000	−0.4%
2001	9033000	+2.0%
2011	9939000	+10.0%

Source: ISTAT 2001

One sixth of the Italian population or about 10 million people live in Lombardy (16.2% of the national population; 2% of the European Union population), making it the second most densely populated region in Italy after Campania with a strong concentration in the Milan metropolitan area and the Alpine foothills areas of the provinces of Varese, Como, Lecco, Monza and Brianza and Bergamo, (1,200 inh./km^2), a lower average density (250 inh./km^2) in the Po valley and the lower Brescia valleys, and much lower densities (less than 60 inh./km^2) in the northern mountain areas and the southern Oltrepò Pavese subregion.

The growth of the regional population was particularly sustained during the 1950s–60s, thanks to a prolonged economic boom, high birth rates, and strong immigration flows (especially from Southern Italy). During the last two decades, Lombardy became the destination of a large number of international immigrants, insomuch that today more than a quarter of all foreign immigrants in Italy lives in this region. As of 2008, the Italian national institute of statistics ISTAT estimated that 815,335 foreign-born immigrants live in Lombardy, equal to 8.4% of the total regional population.

The primary religion is Catholicism; significative religious minorities include Christian Waldenses, Protestants and Orthodox, as well as Jews, Sikh and Muslims.

Economy

View over the business district of Milan: with a metropolitan area of 7.4m people,[9] it is Italy's most important industrial, commercial and financial center.

The gross domestic product in Lombardia (equal to over €328 billion in 2008) accounts for 20% of the total gross domestic product of Italy. When this measure is considered by inhabitant, it results in a value of €31,600 per inhabitant, which is almost 25% higher than the national average of €24,300.[10]

Lombardy's development has been marked by the growth of the services sector since the 1980s, and in particular by the growth of innovative activities in the sector of services to enterprises and in credit and financial services. At the same time, the strong industrial vocation of the region has not suffered from it. Lombardy remains, in fact, the main industrial area of the country. The presence, and development, of a very high number of enterprises belonging to the services sector represents a favourable situation for the improvement of the efficiency of the productive process, as well as for the growth of the regional economy. Lombardy is one of the Four Motors for Europe.

The region can broadly be divided into three areas as regards the productive activity. Milan, where the services sector makes up for 65.3% of the employment; a group of provinces, Varese, Como, Lecco, Bergamo and Brescia, highly industrialised, although in the two latter ones, in the plains, there is also a rich agricultural sector. Finally, in the provinces of Sondrio, Pavia, Cremona, Mantova and Lodi, there is a consistent agricultural activity, and at the same time an above average development of the services sector.

The countryside around Lake Garda.

The productivity of agriculture is enhanced by a well-developed use of fertilizers and the traditional abundance of water, boosted since the Middle Ages by the construction (partly designed by Leonardo da Vinci) of a wide net of irrigation systems. Lower plains are characterized by fodder crops, which are mowed up to eight times a years, cereals (rice, wheat and maize) and sugarbeet. Productions of the higher plains include cereals, vegetables, fruit trees and mulberries. The higher areas, up to the Prealps and Alps sectors of the north, produce fruit and vines. Cattle (with the highest density in Italy), pigs and sheep are raised.

Government and politics

Palazzo Lombardia in Milan, seat of the regional government.

The politics of Lombardy take place in a framework of a presidential representative democracy, whereby the President of the Region (*Presidente della Regione*) is the head of government, and of a pluriform multi-party system. Executive power is exercised by the Regional Government (*Giunta Regionale*). Legislative power is vested in the Regional Council (*Consiglio Regionale*).

The Christian Democrats maintained a large majority of the popular support and the control of the most important cities and provinces since the birth of the Italian Republic until the late 1980s. The Italian Communist Party was a considerable presence only in the eastern and southern parts of Lombardy from the late 1960s to the mid 1980's. Their base however was increasingly eroded by the Italian Socialist Party until, in the early 1990s, the *Mani Pulite* corruption scandal which spread from Milan to the whole of Italy wiped away the old political class almost entirely. This, together with the general disaffection towards Rome's government (considered as favouring excessively the less developed regions of southern Italy in economical matters), led to the sudden growth of the secessionist (later Northern League), that is particularly strong in mountain and rural areas. In recent years, Lombardy stayed as a conservative stronghold, that gave about 60 per cent of its votes to Silvio Berlusconi at the last general election. Notwithstanding, the capital city of Milan landslide elected a new progressive mayor at the 2011 municipal elections.

Culture

Although Lombardy as a region is often identified as merely an economic and industrial powerhouse, it has interesting examples [11] even from the standpoint of cultural and artistic. The many examples range from prehistory to the present day, through the Roman period and the Renaissance and can be found both in museums and churches that enrich cities and towns around the region.

Prehistory

Rock engraves, Nadro.

The rock carvings (some 300,000) left by the ancient Camuni in the Valcamonica depicting animals, people and symbols date back to the period from Neolithic to Middle Ages.

The many artifacts (pottery, personal items and weapons) found in the necropolis near the Lake Maggiore, and Ticino demonstrate the presence of civilization Golasecca who lived in Western Lombardy between the ninth and the fourth century BC.

Museums

Lombardy contains numerous museums (over 330) of different types: ethnographic, historical, technical-scientific, artistic and naturalistic which testify to the historical-cultural and artistic development of the region. Among the most famous are the National Museum of Science and Technology "Leonardo da Vinci" (Milan), the The Last Supper of Leonardo da Vinci (Milan), the Accademia Carrara (Bergamo), the Museum of Santa Giulia (Brescia), the Volta Temple (Como), the Stradivari Museum (Cremona), the Palazzo Te (Mantua), the Museum Sacred Art of the Nativity and the basilica of Santa Maria Assunta at Gandino, the Royal Villa of Monza and many others.

Main sights

Villa Olmo in Como.

- Cathedral of Milan
- Castello Sforzesco
- Basilica di Sant'Ambrogio
- Convent of Sta. Maria delle Grazie
- The Last Supper
- Teatro alla Scala
- Basilica of San Lorenzo
- Basilica of Sant'Eustorgio
- Portinari Chapel
- Brera Gallery
- Certosa di Pavia

- Petroglyphs of Valcamonica
- Bellagio
- Lake of Como
- Lake of Garda
- Industrial village of Crespi d'Adda
- Duomo and Torrazzo in Cremona
- Royal Villa of Monza
- Tempio Civico della Beata Vergine Incoronata in Lodi
- Palazzo Te, Castello di San Giorgio, Basilica di Sant'Andrea and Cathedral in Mantua

The Last Supper (Convent of Sta. Maria delle Grazie, Milan, Italy (1499), by Leonardo da Vinci.

Cuisine

Rice is popular in the region, often found in soups as well as *risotti*, such as "**risotto alla Milanese**", with saffron. In the city of Monza a popular recipe also adds pieces of sausages to the risotto. Regional cheeses include *robiola*, *crescenza*, *taleggio*, *gorgonzola* and *grana padano* (the plains of central and southern Lombardy allow intensive cattle-raising). Butter and cream are used. Single pot dishes, which take less work to prepare, are popular. In Bergamo, Brescia, and Valtellina, polenta is common. In Valtellina, Pizzoccheri too. In Mantua festivals feature *tortelli di zucca* (ravioli with pumpkin filling) accompanied by melted butter and followed by turkey stuffed with chicken or other stewed meats.[12]

Typical dishes

- Polenta (Asino e Polenta, Polenta e Osei, Vunscia Polenta, Polenta e Gorgonzola)
- Pizzoccheri (short tagliatelle made out of buckwheat flour and wheat, laced with butter, green vegetables, garlic, sage, potatoes and onions, all topped with Casera cheese)
- Quartirolo lombardo
- Risotto (alla Milanese)
- Osso buco
- Cotoletta (Cutlet) ("alla Milanese")
- Cassoeula
- Gorgonzola cheese
- Bitto cheese
- Grana Padano cheese
- Panettone
- Lo Spiedo Bresciano – a traditional spit roast consisting of different cuts of meat, butter and sage
- Tortelli di Zucca (Pumpkins filled pasta)
- Sbrisolona cake

A traditional "Cotoletta alla Milanese (Milanese-style cutlet)" served with potatoes.

Wines

- Nebbiolo red
- Bellavista
- Santi
- Nino Negri
- Bonarda Lombardy
- Inferno (Valtellina)
- Grumello (Valtellina)
- Sassella (Valtellina)

Music

The magnificent auditorium of the Teatro Grande in Brescia.

Besides Milan, the region of Lombardy has 11 other provinces, most of them with equally great musical traditions. Bergamo is famous for being the birthplace of Gaetano Donizetti and home of the Teatro Donizetti; Brescia is hosts the impressive 1709 Teatro Grande; Cremona is regarded as the birthplace of the commonly used violin, and is home to several of the most prestigious luthiers in the world, and Mantua was one of the founding and most important cities in 16th and 17th opera and classical music. Other cities such as Lecco, Lodi, Varese and Pavia also have rich musical traditions, but Milan is the hub and centre of the Lombard musical scene. It was the birthplace of Giuseppe Verdi, one of the most famous and influential opera composers of the 19th century, and boasts a variety of acclaimed theatres, such as the Piccolo Teatro and the Teatro Arcimboldi; however, the most famous is the 1778 Teatro alla Scala, one of the most important and prestigious operahouses in the world.

Language

Apart from standardized Italian, Lombard is the local language of Lombardy. Lombard is a member of the Gallo-Italic group within the Romance languages. It is spoken natively in Northern Italy (most of Lombardy and some areas of neighbouring regions, notably the eastern side of Piedmont) and Southern Switzerland (Ticino and Graubünden).

The two main varieties (Western Lombard language and Eastern Lombard language) show differences and are often, but not always, mutually comprehensible. The union of Western Lombard or Insubric, Eastern Lombard and intermediate varieties under the denomination of "Lombard" is a matter of debate, and it has been argued that the two might potentially form separate languages.[13]

Fashion

Lombardy has always been an important centre for silk and textile production, notably the cities of Pavia, Vigevano and Cremona, but Milan is the region's most important centre for clothing and high fashion. In 2009, Milan was regarded as the world fashion capital, even surpassing New York, Paris, Rome and London.[14] Most of the major Italian fashion brands, such as Valentino, Versace, Prada, Armani and Dolce & Gabbana (to name a few), are currently headquartered in the city.

Administrative divisions

Lombardy is divided into 12 provinces:

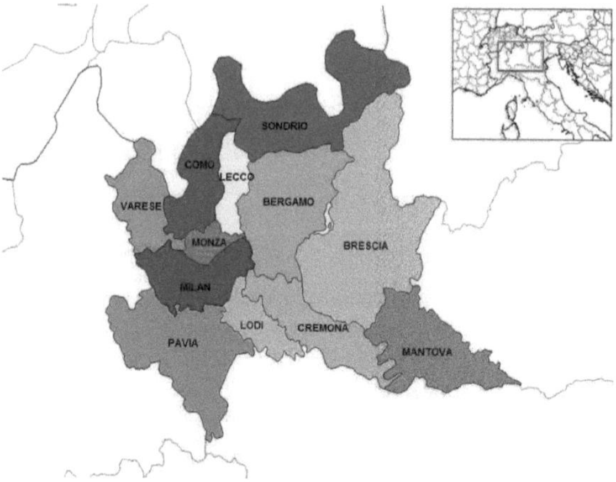

Province	Area (km²)	Population	Density (inh./km²)
Province of Bergamo	2,723	1,070,060	392.9
Province of Brescia	4,784 the biggest	1,223,900	255.8
Province of Como	1,288	582,736	452.4
Province of Cremona	1,772	358,628	202.4
Province of Lecco	816	334,059	409.4
Province of Lodi	782	222,223	284.2
Province of Mantova	2,339	407,983	174.4
Province of Milan	1,575	3,121,832 the biggest	1,982
Province of Monza and Brianza	405 the smallest	840,711	2,075
Province of Pavia	2,965	535,948	180.7
Province of Sondrio	3,212	181,841 the smallest	56.6
Province of Varese	1,199	868,777	724.6

UNESCO World Heritage Sites

- Church and Dominican Convent of Santa Maria delle Grazie with "The Last Supper" by Leonardo da Vinci
- Crespi d'Adda
- Rock Drawings in Valcamonica
- Sacri Monti of Piedmont and Lombardy
- Mantua and Sabbioneta

References

[1] "Eurostat – Tables, Graphs and Maps Interface (TGM) table" (http://epp.eurostat.ec.europa.eu/tgm/table.do?tab=table&init=1&language=en&pcode=tgs00003&plugin=1). Epp.eurostat.ec.europa.eu. 12 August 2011. . Retrieved 16 September 2011.

[2] EUROPA – Press Releases – Regional GDP per inhabitant in 2008 GDP per inhabitant ranged from 28% of the EU27 average in Severozapaden in Bulgaria to 343% in Inner London (http://europa.eu/rapid/pressReleasesAction.do?reference=STAT/11/28&type=HTML)

[3] http://www.regione.lombardia.it/

[4] "EUROPA – Press Releases – Regional GDP per inhabitant in the EU27, GDP per inhabitant in 2006 ranged from 25% of the EU27 average in Nord-Est in Romania to 336% in Inner London" (http://europa.eu/rapid/pressReleasesAction.do?reference=STAT/09/23&format=HTML&aged=0&language=EN&guiLanguage=en). Europa.eu. 19 February 2009. . Retrieved 10 May 2010.

[5] "Euromonitor Internationals Top City Destinations Ranking > Euromonitor archive" (http://www.euromonitor.com/_Euromonitor_Internationals_Top_City_Destinations_Ranking). Euromonitor.com. 12 December 2008. . Retrieved 10 May 2010.

[6] La lingua italiana, i dialetti e le lingue stranieri (http://www.istat.it/salastampa/comunicati/non_calendario/20070420_00/testointegrale.pdf)

[7] "Museo del monastero di Santa Giulia in Brescia" (http://www.santagiulia.info/museo_di_santa_giulia/index.htm). Santagiulia.info. . Retrieved 16 September 2011.

[8] From an italian website a brief history of the pestilence (http://www.storiadimilano.it/Personaggi/Milanesi illustri/mora.htm)

[9] OECD. "Competitive Cities in the Global Economy" (http://213.253.134.43/oecd/pdfs/browseit/0406041E.PDF) (PDF). . Retrieved 30 April 2009.

[10] (http://epp.eurostat.ec.europa.eu/pls/portal/docs/PAGE/PGP_PRD_CAT_PREREL/PGE_CAT_PREREL_YEAR_2008/PGE_CAT_PREREL_YEAR_2008_MONTH_02/1-12022008-EN-AP.PDF)

[11] http://www.lombardiadautore.regione.lombardia.it

[12] Piras, 87.

[13] "Ethnologue report for language code:lmo" (http://www.ethnologue.com/show_language.asp?code=lmo). Ethnologue.com. . Retrieved 16 September 2011.

[14] "The Global Language Monitor » Fashion" (http://www.languagemonitor.com/popular-culture/fashion). Languagemonitor.com. 20 July 2009. . Retrieved 3 January 2010.

Italy

Italian Republic
Repubblica italiana

Anthem: *Il Canto degli Italiani*
The Song of the Italians

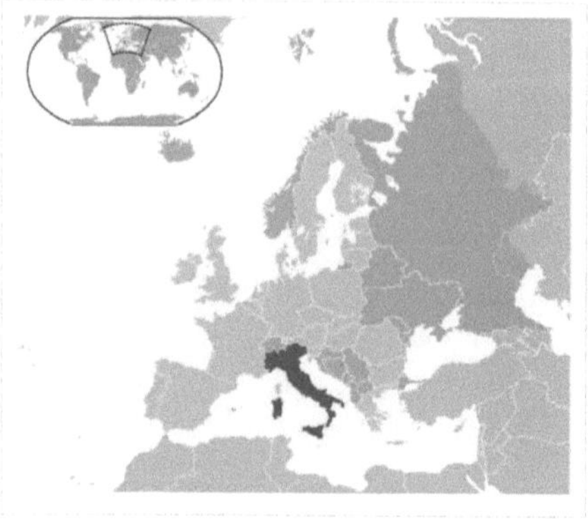

Location of Italy(dark green)

– in Europe(green & dark grey)
– in the European Union(green) — [Legend]

Capital (and largest city)		Rome 41°54′N 12°29′E
Official language(s)		Italian[1]
Demonym		Italian
Government		Unitary parliamentary constitutional republic
-	President	Giorgio Napolitano
-	Prime Minister	Mario Monti
Legislature		Parliament
-	Upper house	Senate of the Republic
-	Lower house	Chamber of Deputies
Formation		
-	Unification	17 March 1861
-	Republic	2 June 1946
Area		

-	Total	301,338 km^2 (71st) 116,346 sq mi
-	Water (%)	2.4
Population		
-	2011 estimate	60,681,514 [2] (23rd)
-	2001 census	56,995,744
-	Density	201.2/km^2 (61st) 521.2/sq mi
GDP (PPP)		2011 estimate
-	Total	$1.847 trillion[3] (10th)
-	Per capita	$30,464[3] (30th)
GDP (nominal)		2011 estimate
-	Total	$2.198 trillion[3] (8th)
-	Per capita	$36,267[3] (24th)
Gini (2006)		32[4]
HDI (2011)		▲0.874[5] (very high) (24th)
Currency		Euro (€)[2] (EUR)
Time zone		CET (UTC+1)
-	Summer (DST)	CEST (UTC+2)
Drives on the		right
ISO 3166 code		IT
Internet TLD		.it[3]
Calling code		39[4]

1	French is co-official in the Aosta Valley; Slovene is co-official in the province of Trieste and the province of Gorizia; German and Ladin are co-official in the province of South Tyrol.
2	Before 2002, the Italian Lira. The euro is accepted in Campione d'Italia, but the official currency there is the Swiss Franc.[6]
3	The .eu domain is also used, as it is shared with other European Union member states.
4	To call Campione d'Italia, it is necessary to use the Swiss code +41.

Italy 🔈 i/ˈɪtəli/ (Italian: *Italia* Italian pronunciation: [iˈtaːlja]), officially the **Italian Republic**[7] or the **Republic of Italy**[8] (Italian: *Repubblica italiana*[9]), is a unitary parliamentary republic in south-central Europe. To the north, it borders France, Switzerland, Austria and Slovenia along the Alps. To the south, it consists of the entirety of the Italian Peninsula, Sicily, Sardinia–the two largest islands in the Mediterranean Sea–and many other smaller islands. The independent states of San Marino and the Vatican City are enclaves within Italy, whilst Campione d'Italia is an Italian exclave in Switzerland. The territory of Italy covers some 301338 km^2 (**unknown operator: u'strong'** sq mi) and is influenced by a temperate seasonal climate. With 60.6 million inhabitants, it is the fifth most populous country in Europe, and the 23rd most populous in the world.

Rome, the capital of Italy, was for centuries a political and religious centre of Western civilisation as the capital of the Roman Empire and site of the Holy See. After the decline of the Roman Empire, Italy endured numerous invasions by foreign peoples, from Germanic tribes such as the Lombards and Ostrogoths, to the Byzantines and later, the Normans, among others. Centuries later, Italy became the birthplace of Maritime republics and the Renaissance,[10] an immensely fruitful intellectual movement that would prove to be integral in shaping the subsequent course of European thought.

Through much of its post-Roman history, Italy was fragmented into numerous city and regional states (such as the Republic of Venice and the Church State), but was unified in 1861,[11] following a tumultuous period in history known as "*Il Risorgimento*" ("The Resurgence"). In the late 19th century, through World War I, and to World War II, Italy possessed a colonial empire, which extended its rule to Libya, Eritrea, Somalia, Ethiopia, Albania, the Dodecanese and a concession in Tianjin, China.[12]

Modern Italy is a democratic republic. It has been ranked as the world's 24th most-developed country[5] and its Quality-of-life index has been ranked in the world's top ten in 2005.[13] Italy enjoys a very high standard of living, and has a high GDP per capita.[14] [15] It is a founding member of what is now the European Union and part of the Eurozone. Italy is also a member of the G8, G20 and NATO. It has the world's third-largest gold reserves, eighth-largest nominal GDP, tenth highest GDP (PPP)[16] and the sixth highest government budget in the world.[17] It is also a member state of the Organisation for Economic Co-operation and Development, the World Trade Organization, the Council of Europe, the Western European Union and the United Nations. Italy has the world's ninth-largest defence budget and shares NATO's nuclear weapons.

Italy plays a prominent role in European and global military, cultural and diplomatic affairs. The country's European political, social and economic influence make it a major regional power.[18] [19] The country has a high public education level and is a highly globalised nation.[20]

Etymology

The assumptions on the etymology of the name "Italia" are very numerous and the corpus of the solutions proposed by historians and linguists is very wide.[21] According to one of the more common explanations, the term *Italia*, from Latin: *Italia*,[22] was borrowed through Greek from the Oscan *Víteliú*, meaning "land of young cattle" (*cf.* Lat *vitulus* "calf", Umb *vitlo* "calf").[23] The bull was a symbol of the southern Italian tribes and was often depicted goring the Roman wolf as a defiant symbol of free Italy during the Social War. Greek historian Dionysius of Halicarnassus states this account together with the legend that Italy was named after Italus,[24] mentioned also by Aristotle[25] and Thucydides.[26]

The name *Italia* originally applied only to a part of what is now Southern Italy–according to Antiochus of Syracuse, the southern portion of the Bruttium peninsula (modern Calabria: province of Reggio, and part of the provinces of Catanzaro and Vibo Valentia). But by his time Oenotria and Italy had become synonymous, and the name also applied to most of Lucania as well. The Greeks gradually came to apply the name "Italia" to a larger region, but it was during the reign of Emperor Augustus (end of the first century BC) that the term was expanded to cover the entire peninsula until the Alps.[27]

History

Prehistory and antiquity

The Colosseum in Rome, built ca. 70 – 80 AD, is considered one of the greatest works of Roman architecture and engineering.

Excavations throughout Italy reveal a Neanderthal presence dating back to the Paleolithic period, some 200,000 years ago,[28] modern Humans arrived about 40,000 years ago. The Ancient peoples of pre-Roman Italy — such as the Umbrians, the Latins (from which the Romans emerged), Volsci, Samnites, the Celts and the Ligures which inhabited northern Italy, and many others — were Indo-European peoples; the main historic peoples of non-Indo-European heritage include the Etruscans, the Elymians and Sicani in Sicily and the prehistoric Sardinians.

Between the 17th and the 11th century BC Mycenaean Greeks established contacts with Italy[29] [30] [31] [32] [33] [34] [35] and in the 8th and 7th centuries BC Greek colonies were established all along the coast of Sicily and the southern part of the Italian Peninsula became known as Magna Graecia. Also the Phoenicians established colonies on the coasts of Sardinia and Sicily.

Ancient Rome was at first a small agricultural community founded c. the 8th century BC, that grew over the course of the centuries into a colossal empire encompassing the whole Mediterranean Sea, in which Ancient Greek and Roman cultures merged into one civilization. This civilization was so influential that parts of it survive in modern law, administration, philosophy and arts, forming the ground that Western civilization is based upon. In a slow decline since the late 2th century AD, the empire finally broke into two parts in 395 AD: the Western Roman Empire and the Eastern Roman Empire. The western part — under the pressure of the Franks, the Vandals, the Huns, the Goths and other populations from Eastern Europe — finally dissolved in 476 AD, when the last western Emperor was deposed by the Barbarian chief Odoacer.

Middle Ages

Italy's Naval Jack, featuring the coats of arms of the four major Maritime Republics. Clockwise from upper left: Venice, Genoa, Pisa, Amalfi.

After the fall of Rome, Italy was conquered by the Germanic Tribe of the Ostrogoths, but in the 6th century the East Roman Emperor Justinian reconquered it. The invasion of another Germanic tribe (the Lombards) late in the same century reduced the Byzantine presence to a strip of land between Ravenna and Rome plus other lands in southern Italy, breaking the unity of the peninsula until 1870.

The Lombard reign of northern and central Italy was absorbed into the Frankish Empire by Charlemagne in the late 8th century. The Frankish kings also helped the formation of the Papal States in central Italy, extending from Rome to Ravenna, although for most of the Middle Ages the Papacy effectively controlled only Latium. The existence of this theocratic state hindered for centuries the unification of the peninsula. Until the 13th century, Italian politics were dominated by the relationship between the German Holy Roman Emperors and the popes, with most of the Italian cities siding for the former (*Ghibellini*) or for the latter (*Guelfi*) depending from momentary convenience.

It was during this vacuum of authority that the Italy saw the rise of a peculiar institution, the medieval commune. In the anarchic conditions that often prevailed in medieval Italian city-states, people organised themselves to restore order and disarm the feuding elites. In the 12th century, a league of comuni, the Lombard League, defeated the

German emperor Frederick Barbarossa, leading to a process granting effective independence to most of northern and central Italian cities. Despite the devastation of the numerous wars, Italy maintained, especially in the north and center, a relatively developed urban civilization.

During the same period, Italy saw the rise of numerous Maritime Republics, the most notable being Venice, Genoa, Pisa and Amalfi. Heavily involved in the Crusades, they took advantage of political and trading opportunities. Venice and Genoa soon became Europe's main gateways to trade with the East, establishing colonies as far as the Black Sea and often controlling most of the trade with the Byzantine Empire and the Islamic Mediterranean world. The county of Savoy expanded its territory into the peninsula in the late Middle Ages, while Florence developed into a highly organized commercial and financial city-state, becoming for many centuries the European capital of silk, wool, banking and jewelry.

In the south, Byzantine Sicily had become an Islamic emirate in the 9th century, thriving until the Italo-Normans conquered it in the late 11th century together with most of the Lombard and Byzantine states of southern Italy. Through a complex series of events, southern Italy developed as a unified kingdom, first under the House of Hohenstaufen, then under the Capetian House of Anjou and, from the 15th century, the house of Aragon (although Sicily was a separate Aragonese kingdom from the late 13th to the 15th century). In Sardinia, the former Byzantine provinces became independent states known as giudicati, although most of the island was under Genoese or Pisan control until the Aragonese conquered it in the 15th century.

Early Modern

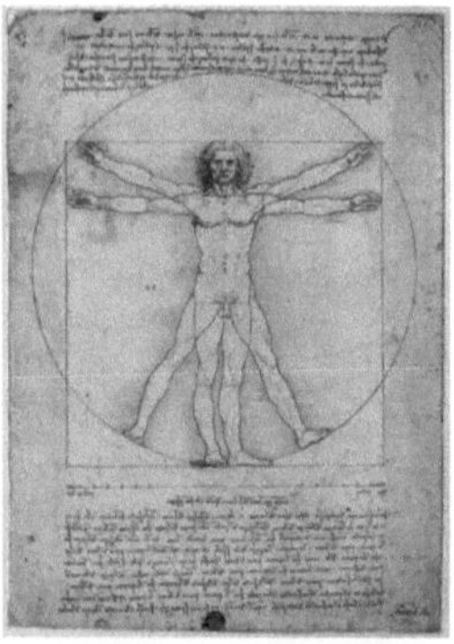

The Vitruvian man by Leonardo da Vinci, representing the ideal human proportions as described by Roman architect Vitruvius, is a quintessential masterpiece of the Renaissance.

The Black Death pandemic in 1348 left its mark on Italy by killing one third of the population.[36] [37] However, the recovery from the disaster of the Black Death led to a resurgence of cities, trade and economy which greatly stimulated the successive phases of Humanism and Renaissance, cultural movements both born in the peninsula, and later spread in Europe.

In the 14th and 15th centuries, Northern and upper Central Italy were divided into a number of warring city-states, the rest of the peninsula being occupied by the larger Papal States and Naples. The strongest among these city-states annexed the surrounding territories giving birth to the Signorie, regional states led by merchant families which founded local dynasties. Dominated by merchant oligarchies, they enjoyed a relative freedom and nurtured academic and artistic advancement. Warfare between the states was common, invasion from outside Italy confined to intermittent sorties of Holy Roman Emperors. These wars were primarily fought by armies of mercenaries known as *condottieri*, bands of soldiers drawn from around Europe, but especially Germany and Switzerland, led largely by Italian captains.[38]

Decades of fighting eventually saw Florence, Milan and Venice emerge as the dominant players that agreed to the Peace of Lodi in 1454, which saw relative calm brought to the region for the first time in centuries. This peace would hold for the next forty years, and Venice's unquestioned hegemony over the sea also led to unprecedented peace for much of the rest of the 15th century. The Italian Renaissance peaked in the mid-16th century as foreign invasions plunged the region into the turmoil of the Italian Wars. However, the ideas and ideals of the Renaissance endured and even spread into the rest of Europe, setting off the Northern Renaissance, and the English Renaissance.

In the meantime, the discovery of the Americas, the new routes to Asia discovered by the Portuguese and the rise of the Ottoman Empire — all factors which eroded the traditional Italian dominance in trade with the East — started the economic decline of the peninsula.

Following the Italian Wars (1494 to 1559), Italy saw a long period of relative peace, first under Habsburg Spain (1559 to 1713) and then under Habsburg Austria (1713 to 1796). The plague repeatedly returned to haunt Italy throughout the 14th to 17th centuries. In the first half of the 17th century, a plague claimed some 1.7 million victims, or about 14% of Italy's population.[39] As Spain declined in the 17th century, so did its Italian possessions in Naples, Sicily, Sardinia, and Milan. Southern Italy was impoverished, stagnant, and cut off from the mainstream of events in Europe.[40] Despite that, Italy kept making its contribution to the European culture, giving birth to the Baroque Style.

The triumph of Napoleon at the Battle of Marengo placed Italy under French control and paved him the way to become Emperor.

In the 18th century, as a result of the War of Spanish Succession, Austria replaced Spain as the dominant foreign power, while the House of Savoy emerged as a major regional power expanding to Piedmont and Sardinia. In this century, the ideas of the Enlightenment influenced the Italian rulers, paving the way to reforms which started an economic recovery in northern Italy and Tuscany.

During the Napoleonic Wars, the northern and central parts of the country were invaded and later partly annexed to the Empire and partly reorganized as a new Kingdom of Italy — essentially a client state of the French Empire —[41] while the southern half of the peninsula was administered by Joachim Murat, Napoleon's brother-in-law, who was crowned as King of Naples. The 1814 Congress of Vienna restored the situation of the late 18th century, but the ideals of the French Revolution could not be eradicated.

Italian unification and Liberal Italy

The legendary "handshake of Teano" between Giuseppe Garibaldi and Victor Emmanuel II: on 26 October 1860, General Garibaldi sacrificed republican hopes for the sake of Italian unity under a monarchy.

The creation of the Kingdom of Italy was the result of efforts by Italian nationalists and monarchists loyal to the House of Savoy to establish a united state encompassing the entire Italian Peninsula. In the context of the 1848 liberal revolutions that swept through Europe, an unsuccessful war was declared on Austria. The Kingdom of Sardinia again attacked the Austrian Empire in the Second Italian War of Independence of 1859, with the aid of France, resulting in liberating Lombardy. In 1860-61, Giuseppe Garibaldi led the drive for unification in Naples and Sicily,[42] allowing the Sardinian government led by the Count of Cavour to declare a united Italian kingdom on 17 March 1861. In 1866, Victor Emmanuel II allied with Prussia during the Austro-Prussian War, waging the Third Italian War of Independence which allowed Italy to annex Venetia. Finally, as France during the disastrous Franco-Prussian War of 1870 abandoned its garrisons in Rome, the Savoy rushed to fill the power gap by taking over the Papal States.

Italian infantrymen in 1916. More than 650,000 Italian soldiers lost their lives on the battlefields of World War I.

The Sardinian Albertine Statute of 1848, extended to the whole Kingdom of Italy in 1861, provided for basic freedoms, but electoral laws excluded the non-propertied and uneducated classes from voting. The government of the new kingdom took place in a framework of parliamentary constitutional monarchy dominated by liberal forces. In 1913, male universal suffrage was adopted. As Northern Italy quickly industrialized, the South and rural areas of North remained underdeveloped and overpopulated, forcing millions of people to migrate abroad, while the Italian Socialist Party constantly increased in strength, challenging the traditional liberal and conservative establishment.

Starting from the last two decades of the 19th century, Italy developed into a colonial power by forcing Somalia, Eritrea and later Libya and the Dodecanese under its rule.[43] During World War I, Italy at first stayed neutral, but in 1915 signed the Treaty of London, entering the Entente on the promise of receiving Trento, Trieste, Gorizia and Gradisca, Istria and northern Dalmatia from the Austro-Hungarian Empire — as well as parts of the Ottoman Empire. During the war, more than 650,000 Italian soldiers died,[44] and the economy collapsed. Under the Peace Treaties of Saint-Germain, Rapallo and Rome, Italy obtained most of the promised territories, including the Hungarian harbour of Fiume, but not Dalmatia (except Zara), allowing nationalists to define the victory as "mutilated".

Fascist regime

Benito Mussolini and Adolf Hitler in 1940.

The turbulence that followed the devastation of World War I, inspired by the Russian Revolution, led to turmoil and anarchy. The liberal establishment, fearing a socialist revolution, started to endorse the small National Fascist Party, led by Benito Mussolini. In October 1922 the fascists attempted a coup (the "March on Rome"), supported by king Victor Emmanuel III. Over the next few years, Mussolini banned all political parties and curtailed personal liberties, thus forming a dictatorship. In 1935, Mussolini invaded Ethiopia, resulting in an international alienation and leading to Italy's withdrawal from the League of Nations. Consequently, Italy allied with Nazi Germany and Empire of Japan and strongly supported Franco in the Spanish civil war. In 1939, Italy occupied Albania, a *de facto* protectorate for decades, and entered World War II in June 1940 on the side of the Axis powers. Mussolini, wanting a quick victory like Hitler's Blitzkriegs in Poland and France, invaded Greece in October 1940, but was forced to accept a humiliating stalemate after a few months. At the same time, Italy, after initially conquering British Somalia and parts of Egypt, saw an allied counter-attack lead to the loss of all possessions in the Horn of Africa and in North Africa.

Italy was then invaded by the Allies in July 1943, leading to the collapse of the Fascist regime and the fall of Mussolini. In September 1943, Italy surrendered. The country remained a battlefield for the rest of the war, as the allies were moving up from the south as the north was the base for loyalist Italian fascist and German Nazi forces, fought also by the Italian resistance movement. The hostilities ended on 2 May 1945. Nearly half a million Italians (including civilians) died in the conflict,[45] and the Italian economy had been all but destroyed; per capita income in 1944 was at its lowest point since the beginning of the 20th century.[46]

Italian Republic

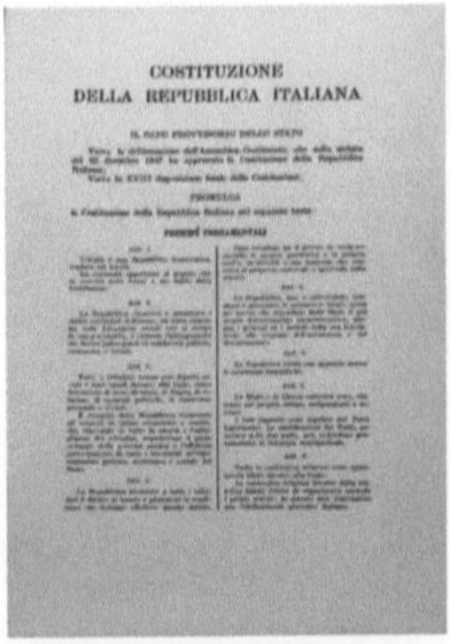
COSTITUZIONE
DELLA REPUBBLICA ITALIANA

The Constitution of Italy, come into force in 1948.

Italy became a republic after a referendum[47] held on 2 June 1946, a day celebrated since as Republic Day. This was also the first time that Italian women were entitled to vote.[48] Victor Emmanuel III's son, Umberto II, was forced to abdicate and exiled. The Republican Constitution was approved on 1 January 1948. Under the Paris Peace Treaties of 1947, the eastern border area was lost to Yugoslavia, and, later, the Free Territory of Trieste was divided between the two states. Fears in the Italian electorate of a possible Communist takeover proved crucial for the first universal suffrage electoral outcome on 18 April 1948, when the Christian Democrats, under the leadership of Alcide De Gasperi, obtained a landslide victory. Consequently, in 1949 Italy became a member of NATO. The Marshall Plan helped to revive the Italian economy which, until the late 1960s, enjoyed a period of sustained economic growth commonly called the "Economic Miracle". In 1957, Italy was a founding member of the European Economic Community (EEC), which became the European Union (EU) in 1993.

The 1957 Treaties of Rome signing ceremony. Italy is a founding member of the European Union.

From the late 1960s until the early 1980s, the country experienced the Years of Lead, a period characterized by economic crisis (especially after the 1973 oil crisis), widespread social conflicts and terrorist massacres carried out by opposing extremist groups, with the alleged involvement of US intelligence.[49] [50] [51] The Years of Lead culminated in the assassination of the Christian Democrat leader Aldo Moro in 1978, an event that deeply affected the whole country. In the 1980s, for the first time since 1945, two governments were led by non-Christian-Democrat premiers: one liberal (Giovanni Spadolini) and one socialist (Bettino Craxi); the Christian Democrats remained, however, the main government party. During Craxi's government, the economy recovered and Italy became the world's fifth largest industrial nation, gaining entry into the G7 Group. However, as a result of his spending policies, the Italian national debt skyrocketed during the Craxi era, soon passing 100% of the GDP.

In the early 1990s, Italy faced significant challenges, as voters — disenchanted with political paralysis, massive public debt and the extensive corruption system (known as *Tangentopoli*) uncovered by the 'Clean Hands' investigation — demanded radical reforms. The scandals involved all major parties, but especially those in the government coalition: the Christian Democrats, who ruled for almost 50 years, underwent a severe crisis and eventually disbanded, splitting up into several factions. The Communists reorganized as a social-democratic force. During the 1990s and the 2000s (decade), centre-right (dominated by media magnate Silvio Berlusconi) and centre-left coalitions alternatively governed the country, which entered a prolonged period of economic stagnation.

Geography

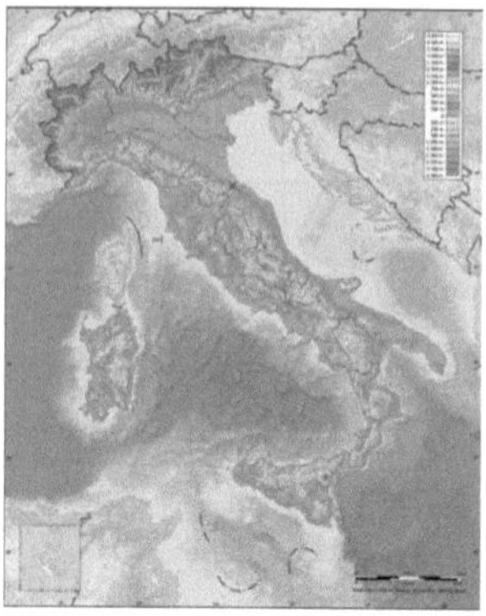

Topographic map of Italy.

Italy is located in Southern Europe and comprises the boot-shaped Italian Peninsula and a number of islands including the two largest, Sicily and Sardinia. It lies between latitudes 35° and 47° N, and longitudes 6° and 19° E. Although the country comprises the Italian peninsula and most of the southern Alpine basin, some of Italy's territory extends beyond the Alpine basin and some islands are located outside the Eurasian continental shelf. These territories are the *comuni* of: Livigno, Sexten, Innichen, Toblach (in part), Chiusaforte, Tarvisio, Graun im Vinschgau (in part), which are all part of the Danube's drainage basin, while the Val di Lei constitutes part of the Rhine's basin and the islands of Lampedusa and Lampione are on the African continental shelf.

The country's total area is 301,230 km², of which 294,020 km² is land and 7,210 km² is water. Including the islands, Italy has a coastline and border of 7,600 km on the Adriatic, Ionian, Tyrrhenian seas (740 km), and borders shared with France (488 km), Austria (430 km), Slovenia (232 km) and Switzerland; San Marino (39 km) and Vatican City (3.2 km), both enclaves, account for the remainder.

Mont Blanc is the highest point in Italy and the European Union.

The Apennine Mountains form the peninsula's backbone and the Alps form its northern boundary, where Italy's highest point is located on Mont Blanc (4,810 m/15,782 ft).[52] The Po, Italy's longest river (652 km/405 mi), flows from the Alps on the western border with France and crosses the Padan plain on its way to the Adriatic Sea. The five largest lakes are, in order of diminishing size:[53] Garda (367.94 km^2/**unknown operator: u'strong'** sq mi), Maggiore (212.51 km^2/**unknown operator: u'strong'** sq mi, shared with Switzerland), Como (145.9 km^2/**unknown operator: u'strong'** sq mi), Trasimeno (124.29 km^2/**unknown operator: u'strong'** sq mi) and Bolsena (113.55 km^2/**unknown operator: u'strong'** sq mi).

The country is situated at the meeting point of the Eurasian Plate and the African Plate, leading to considerable seismic and volcanic activity. There are 14 volcanoes in Italy, three of which are active: Etna (the traditional site of Vulcan's smithy), Stromboli and Vesuvius. Vesuvius is the only active volcano in mainland Europe and is most famous for the destruction of Pompeii and Herculanum. Several islands and hills have been created by volcanic activity, and there is still a large active caldera, the Campi Flegrei north-west of Naples.

Environment

After its quick industrial growth, Italy took a long time to confront its environmental problems. After several improvements, it now ranks 84th in the world for ecological sustainability.[54] National parks cover about five percent of the country.[55] In the last decade, Italy has become one of the world's leading producers of renewable energy, ranking as the world's fourth largest holder of installed solar energy capacity [56] [57] and the sixth largest holder of wind power capacity in 2010.[58] Renewable energies now make up about 12% of the total primary and final energy consumption in Italy, with a future target share set at 17% for the year 2020.[59]

Map of national parks in Italy.

Asinara island with the Aragonese Torre della Pelosa (16th century), Sardinia.

However, air pollution remains a severe problem, especially in the industrialised north, reaching the tenth highest level worldwide of industrial carbon dioxide emissions in the 1990s.[60] Italy is the twelfth largest carbon dioxide producer.[61] [62] Extensive traffic and congestion in the largest metropolitan areas continue to cause severe environmental and health issues, even if smog levels have decreased dramatically since the 1970s and 1980s, and the presence of smog is becoming an increasingly rarer phenomenon and levels of sulphur dioxide are decreasing.[63]

Many watercourses and coastal stretches have also been contaminated by industrial and agricultural activity, while due to rising water levels, Venice has been regularly flooded throughout recent years. Waste from industrial activity is not always disposed of by legal means and has led to permanent health effects on inhabitants of affected areas, as in the case of the Seveso disaster. The country has also operated several nuclear reactors between 1963 and 1990 but, after the Chernobyl disaster and a referendum on the issue the nuclear program was terminated, a decision that was overturned by the government in 2008, planning up to four French nuclear power plants. This was in turn struck down by a referendum following the Fukushima Daiichi nuclear disaster.[64] Deforestation, illegal building developments and poor land-management policies have led to significant erosion all over Italy's mountainous regions, leading to major ecological disasters like the 1963 Vajont Dam flood, the 1998 Sarno[65] and 2009 Messina mudslides.

Climate

The coastal areas of Liguria have a Mediterranean climate.

The climate of Italy is highly diverse and can be quite different from the stereotypical Mediterranean climate. Most of the inland northern regions of Italy, for example Piedmont, Lombardy and Emilia-Romagna, have a climate variously described as humid continental or temperate. The climate of the "Po valley region [is] continental ... with harsh winters and hot summers".[66] [67] The coastal areas of Liguria and most of the peninsula south of Florence generally fit the Mediterranean stereotype (Köppen climate classification Csa). Conditions on peninsular coastal areas can be very different from the interior's higher ground and valleys, particularly during the winter months when the higher altitudes tend to be cold, wet, and often snowy. The coastal regions have mild winters and warm and generally dry summers, although lowland valleys can be quite hot in summer.

Administrative divisions

Italy is subdivided into 20 regions (*regioni*, singular *regione*), five of these regions having a special autonomous status that enables them to enact legislation on some of their local matters. The country is further divided into 110 provinces (*province*) and 8,100 municipalities (*comuni*). There are also 15 metropolitan cities (*città metropolitane*), established in 2009, but this administrative division is not yet operational.

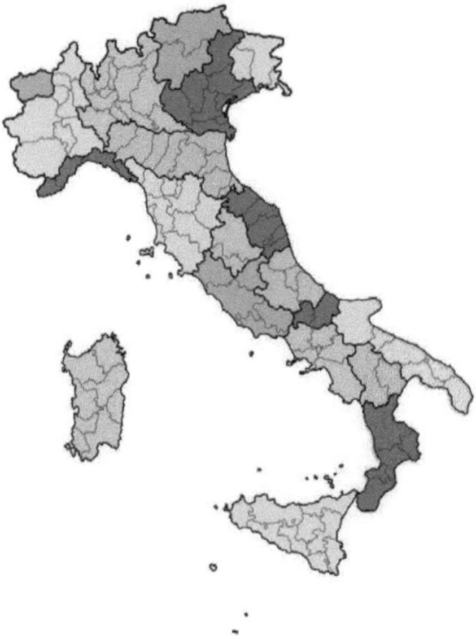

Apulia

Basilicata

Calabria

Sicily

Molise

Campania

Abruzzo

Lazio

Umbria

Marche

Tuscany

Sardinia

Emilia-Romagna

Liguria

Piedmont

Friuli
Venezia Giulia

Aosta
Valley

Trentino
Alto Adige
Südtirol

Veneto

Lombardy

Adriatic Sea

Ionian Sea

Mediterranean Sea

Tyrrhenian Sea

Ligurian Sea

Region[68]	Capital	Area (km²)	Area (sq mi)	Population
Abruzzo	L'Aquila	10,763	4,156	1,342,177
Aosta Valley	Aosta	3,263	1,260	128,129
Apulia	Bari	19,358	7,474	4,090,577
Basilicata	Potenza	9,995	3,859	587,680
Calabria	Catanzaro	15,080	5,822	2,011,537
Campania	Naples	13,590	5,247	5,833,131
Emilia-Romagna	Bologna	22,446	8,666	4,429,766
Friuli-Venezia Giulia	Trieste	7,858	3,034	1,235,761
Lazio	Rome	17,236	6,655	5,724,365
Liguria	Genoa	5,422	2,093	1,616,993
Lombardy	Milan	23,844	9,206	9,909,348
Marche	Ancona	9,366	3,616	1,564,886
Molise	Campobasso	4,438	1,713	319,834
Piedmont	Turin	25,402	9,808	4,456,532

Sardinia	Cagliari	24,090	9,301	1,675,286
Sicily	Palermo	25,711	9,927	5,050,486
Tuscany	Florence	22,993	8,878	3,749,074
Trentino-Alto Adige/Südtirol	Trento	13,607	5,254	1,036,639
Umbria	Perugia	8,456	3,265	906,675
Veneto	Venice	18,399	7,104	4,936,197

Politics

Italy has been a unitary parliamentary republic since 2 June 1946, when the monarchy was abolished by a constitutional referendum. The President of the Italian Republic (*Presidente della Repubblica*), currently Giorgio Napolitano since 2006, is Italy's head of state. The President is elected for a single seven years mandate by the Parliament in joint session. Italy has a written democratic constitution, resulting from the work of a Constituent Assembly formed by the representatives of all the anti-fascist forces that contributed to the defeat of Nazi and Fascist forces during the Civil War.[69]

Giorgio Napolitano, 11th President of the Italian Republic.

Government

Italy has a parliamentary government based on a proportional voting system. The Parliament of Italy is perfectly bicameral: the two houses, the Chamber of Deputies (that meets in Palazzo Montecitorio) and the Senate of the Republic (that meets in Palazzo Madama), have the same powers. The Prime Minister, officially President of the Council of Ministers (*Presidente del Consiglio dei Ministri*), is Italy's head of government. The Prime Minister and the cabinet are appointed by the President of the Republic, but must pass a vote of confidence in Parliament to become in office. While the office is similar to those in most other parliamentary systems, the Italian prime minister has less authority than some of his counterparts. The prime minister is not authorized to request the dissolution of Parliament or dismiss ministers (that are exclusive prerogatives of the President of the Republic) and must receive a vote of approval from the Council of Ministers—which holds effective executive power—to execute most political activities.

Palazzo Montecitorio, seat of the Italian Chamber of Deputies.

Silvio Berlusconi, from 8 May 2008 to his resignation on 12 November 2011, was Prime Minister, leading a center-right coalition. The Italy's four major political parties are the People of Freedom, the Democratic Party, the Northern League and the Italy of Values. During the 2008 general elections these four parties won 590 out of 630 seats available in the Chamber of Deputies and 308 out of 315 seats available in the Senate of the Republic. Most of the remaining seats were won by minor parties that only contest election in one part of Italy, like the South Tyrolean People's Party and the Movement for Autonomies. However, during the last 3 years, a so called "Third Pole" emerged, merging the Christian Democrats of UDC with some dissident MPs coming from Mr. Berlusconi's cabinet. A peculiarity of the Italian Parliament is the representation given to Italian citizens permanently living abroad (about 3.6 million people): 12 Deputies and 6 Senators elected in four distinct overseas constituencies. In addition, the Italian Senate is characterized also by a small number of senators for life, appointed by the President "for outstanding patriotic merits in the social, scientific, artistic or literary field". Former Presidents of the Republic are *ex officio* life senators.

Law and criminal justice

The Supreme Court of Cassation.

The Italian judicial system is based on Roman law modified by the Napoleonic code and later statutes. The Supreme Court of Cassation is the highest court in Italy for both criminal and civil appeal cases. The Constitutional Court of Italy (*Corte Costituzionale*) rules on the conformity of laws with the Constitution and is a post–World War II innovation. Since their appearance in the middle of the 19th century, Italian organized crime and criminal organizations have infiltrated the social and economic life of many regions in Southern Italy, the most notorious of which being the Sicilian Mafia, which would later expand into some foreign countries including the United States. The Mafia receipts may reach 9%[70] [71] of Italy's GDP.[72] A 2009 report identified 610 comuni which have a strong Mafia presence, where 13 million Italians live and 14.6% of the Italian GDP is produced.[73] [74] The Calabrian 'Ndrangheta, nowadays probably the most powerful crime syndicate of Italy, accounts alone for 3% of the country's GDP.[75] However, at 0.013 per 1,000 people, Italy has only the 47th highest murder rate[76] (in a group of 62 countries) and the 43rd highest number of rapes per 1,000 people in the world (in a group of 65 countries), relatively low figures among developed countries.

Foreign relations

US President Barack Obama and Giorgio Napolitano in Rome.

Italy is a founding member of the European Community, now the European Union (EU), and of the North Atlantic Treaty Organization (NATO). Italy was admitted to the United Nations in 1955, and it is a member and strong supporter of a wide number of international organizations, such as the Organisation for Economic Co-operation and Development (OECD), the General Agreement on Tariffs and Trade/World Trade Organization (GATT/WTO), the Organization for Security and Co-operation in Europe (OSCE), the Council of Europe, and the Central European Initiative. Its recent turns in the rotating presidency of international organisations include the Conference for Security and Co-operation in Europe (CSCE), the forerunner of the OSCE, in 1994; G8; and the EU in 2009 and from July to December 2003.

Italy strongly supports multilateral international politics, endorsing the United Nations and its international security activities. Italy deployed troops in support of UN peacekeeping missions in Somalia, Mozambique, and East Timor and provides support for NATO and UN operations in Bosnia, Kosovo and Albania. Italy deployed over 2,000 troops in Afghanistan in support of Operation Enduring Freedom (OEF) from February 2003. Italy still supports international efforts to reconstruct and stabilize Iraq, but it has withdrawn its military contingent of some 3,200 troops as of November 2006, maintaining only humanitarian operators and other civilian personnel. In August 2006 Italy deployed about 2,450 troops in Lebanon for the United Nations' peacekeeping mission UNIFIL.[77]

Military

The new Italian aircraft carrier Cavour (550) (*MM Cavour*).

The Italian Army, Navy, Air Force and Gendarmerie collectively form the Italian armed forces, under the command of the Supreme Defence Council, presided over by the President of the Italian Republic. From 1999, military service is voluntary.[78] In 2010, the Italian military had 293,202 personnel on active duty,[79] of which 114,778 in the national gendarmerie.[80] Total Italian military spending in 2010 ranked tenth in the world, standing at $35.8 billion, equal to 1.7% of national GDP. As part of NATO's nuclear sharing strategy Italy also hosts 90 United States nuclear bombs, located in the Ghedi and Aviano air bases.[81]

The Italian Army is the national ground defense force, numbering 109,703 in 2008. Its best-known combat vehicles are the Dardo infantry fighting vehicle, the Centauro tank destroyer and the Ariete tank, and among its aircraft the Mangusta attack helicopter, recently deployed in UN missions. It also has at its disposal a large number of Leopard 1 and M113 armored vehicles.

A Eurofighter Typhoon operated by the Italian Air Force.

The Italian Navy in 2008 had 35,200 active personnel with 85 commissioned ships and 123 aircraft.[82] It is now equipping itself with a bigger aircraft carrier, (the *Cavour*), new destroyers, submarines and multipurpose frigates. In modern times the Italian Navy, being a member of the North Atlantic Treaty Organization (NATO), has taken part in many coalition peacekeeping operations around the world.

The Italian Air Force in 2008 had a strength of 43,882 and operated 585 aircraft, including 219 combat jets and 114 helicopters. As a stopgap and as replacement for leased Tornado ADV interceptors, the AMI has leased 30 F-16A Block 15 ADF and four F-16B Block 10 Fighting Falcons, with an option for more. The coming years also will see the introduction of 121 EF2000 Eurofighter Typhoons, replacing the leased F-16 Fighting Falcons. Further updates are foreseen in the Tornado IDS/IDT and AMX fleets. A transport capability is guaranteed by a fleet of 22 C-130Js and Aeritalia G.222s of which 12 are being replaced with the newly developed G.222 variant called the C-27J Spartan.

An autonomous corps of the military, the Carabinieri are the gendarmerie and military police of Italy, policing the military and civilian population alongside Italy's other police forces. While the different branches of the Carabinieri report to separate ministries for each of their individual functions, the corps reports to the Ministry of Internal Affairs when maintaining public order and security.[83]

Economy

The new Ferrari F12berlinetta. Italy is the world's 7th largest exporter of goods.

Italy has a free market economy characterized by high per capita GDP and low unemployment rates. In 2010, it was the eighth-largest economy in the world and the fourth-largest in Europe in terms of nominal GDP,[84] and the tenth-largest economy in the world and fifth-largest in Europe in terms of PPP.[85] It is a founding member of the G8, the Eurozone and the OECD.

After World War II, Italy was rapidly transformed from an agriculture based economy into one of the world's most industrialized nations[86] and a leading country in world trade and exports. It is a developed country, with the world's 8th highest quality of life[13] and the 23rd Human Development Index. In spite of the recent global economic crisis, Italian per capita GDP at purchasing power parity remains approximately equal to the EU average,[87] while the unemployment rate (8.5%) stands as one of the EU's lowest.[88] The country is well known for its influential and innovative business economic sector,[89] an industrious and competitive agricultural sector[89] (Italy is the world's largest wine producer),[90] and for its creative and high-quality automobile, industrial, appliance and fashion design.[89]

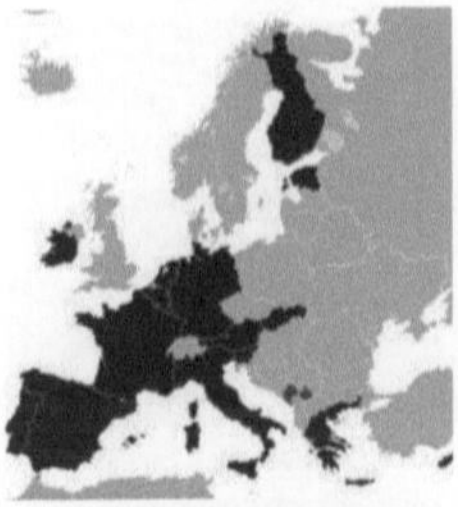
Italy is part of a monetary union, the Eurozone (dark blue), and of the EU single market.

Italy has a smaller number of global multinational corporations than other economies of comparable size, but there is a large number of small and medium-sized enterprises, notoriously clustered in several industrial districts, which are the backbone of the Italian industry. This has produced a manufacturing sector often focused on the export of niche market and luxury products, that if on one side is less capable to compete on the quantity, on the other side is more capable of facing the competition from China and other emerging Asian economies based on lower labour costs, with higher quality products.[91] The country was the world's 7th largest exporter in 2009.[92] Italy's closest trade ties are with the other countries of the European Union, with whom it conducts about 59% of its total trade. Its largest EU trade partners, in order of market share, are Germany (12.9%), France (11.4%), and Spain (7.4%).[93] Finally, tourism is one of the fastest growing and profitable sectors of the national economy: with 43.6 million international tourist arrivals and total receipts estimated at $38.8 billion in 2010, Italy is both the fifth most visited country and highest tourism earner in the world.[94]

Vineyards in the hilly area of Langhe, Piedmont. Italy is the world's top wine producer.

Despite these important achievements, the Italian economy today suffers from many and relevant problems. After a strong GDP growth of 5–6% per year from the 1950s to the early 1970s,[95] and a progressive slowdown in the 1980s and 1990s, the last decade's average annual growth rates poorly performed at 1.23% in comparison to an average EU annual growth rate of 2.28%.[96] The stagnation in economic growth, and the political efforts to revive it with massive government spending from the 1980s onwards, eventually produced a severe rise in public debt. According to the EU's statistics body Eurostat, Italian public debt stood at 116% of GDP in 2010, ranking as the second biggest debt ratio after Greece (with 126.8%).[97] However, the biggest chunk of Italian public debt is owned by national subjects, a major difference between Italy and Greece.[98] In addition, Italian living standards have a considerable north-south divide. The average GDP per capita in the north exceeds by far the EU average, whilst many regions of Southern Italy are dramatically below.[99] Italy has often been referred the *sick man of Europe*,[100] [101] characterised by economic stagnation, political instability and problems in pursuing reform programs.

More specifically, Italy suffers from structural weaknesses due to its geographical conformation and the lack of raw materials and energy resources: in 2006 the country imported more than 86% of its total energy consumption (99.7% of the solid fuels, 92.5% of oil, 91.2% of natural gas and 15% of electricity).[102] [103] The Italian economy is weakened by the lack of infrastructure development, market reforms and research investment, and also high public deficit.[89] In the Index of Economic Freedom 2008, the country ranked 64th in the world and 29th in Europe, the lowest rating in the Eurozone. Italy still receives development assistance from the European Union every year. Between 2000 and 2006, Italy received €27.4 billion from the EU.[104] The country has an inefficient state bureaucracy, low property rights protection and high levels of corruption, heavy taxation and public spending that accounts for about half of the national GDP.[105] In addition, the most recent data show that Italy's spending in R&D

in 2006 was equal to 1.14% of GDP, below the EU average of 1.84% and the Lisbon Strategy target of devoting 3% of GDP to research and development activities.[106] According to the Confesercenti, a major business association in Italy, organized crime in Italy represented the "biggest segment of the Italian economy", accounting for €90 billion in receipts and 7% of Italy's GDP.[107]

Infrastructure

Frecciarossa high speed trains, capable of 355 km/h (**unknown operator: u'strong'** mph).[108] [109] [110]

In 2004 the transport sector in Italy generated a turnover of about 119.4 billion euros, employing 935,700 persons in 153,700 enterprises. Regarding the national road network, in 2002 there were **unknown operator: u',' unknown operator: u',' unknown operator: u','** (**unknown operator: u'strong' unknown operator: u','**mi) of serviceable roads in Italy, including **unknown operator: u',' unknown operator: u',' unknown operator: u','** (**unknown operator: u'strong' unknown operator: u','**mi) of motorways, state-owned but privately operated by Atlantia. In 2005, about 34,667,000 passenger cars (590 cars per 1,000 people) and 4,015,000 goods vehicles circulated on the national road network.[111] The national railway network, state-owned and operated by Ferrovie dello Stato, in 2003 totalled **unknown operator: u',' unknown operator: u',' unknown operator: u','** (**unknown operator: u'strong' unknown operator: u','**mi) of which 69% is electrified, and on which 4,937 locomotives and railcars circulated. The national inland waterways network comprised **unknown operator: u',' unknown operator: u',' unknown operator: u','** (**unknown operator: u'strong' unknown operator: u','**mi) of navigable rivers and channels in 2002. In 2004 there were approximately 30 main airports (including the two hubs of Malpensa International in Milan and Leonardo Da Vinci International in Rome) and 43 major seaports (including the seaport of Genoa, the country's largest and second largest in the Mediterranean Sea). In 2005 Italy maintained a civilian air fleet of about 389,000 units and a merchant fleet of 581 ships.[111]

Demographics

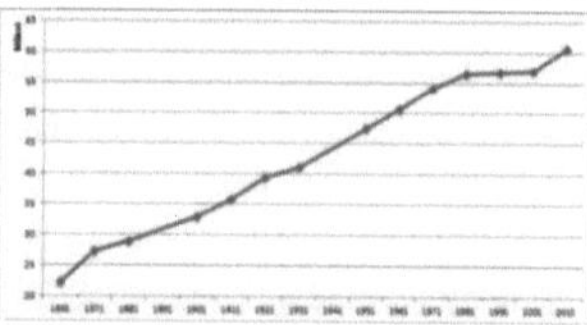

Population 1861–2008. Number of inhabitants.

Italy has 60,626,442 inhabitants according to 1/1/2011 municipal records (*Anagrafe*).[112] Its population density, at 201/km² (520/sq. mile), is higher than that of most Western European countries. However the distribution of the population is widely uneven. The most densely populated areas are the Po Valley (that accounts for almost a half of the national population) and the metropolitan areas of Rome and Naples, while vast regions such as the Alps and Appennines highlands, the plateaus of Basilicata and the island of Sardinia are very sparsely populated.

The population of Italy almost doubled during the 20th century, but the pattern of growth was extremely uneven due to large-scale internal migration from the rural South to the industrial cities of the North, a phenomenon which happened as a consequence of the Italian economic miracle of the 1950-1960s. In addition, after centuries of net emigration, from the 1980s Italy has experienced large-scale immigration for the first time in modern history. According to the Italian government, there were 4,570,317 foreign residents in Italy as of January 2011.[113]

High fertility and birth rates persisted until the 1970s, after which they start to dramatically decline, leading to rapid population aging. At the end of the 2000s (decade), one in five Italians was over 65 years old.[114] However, thanks mainly to the massive immigration of the last two decades, in recent years Italy experienced a significant growth in birth rates.[115] The total fertility rate has also climbed from an all-time low of 1.18 children per woman in 1995 to 1.41 in 2008.[116]

Ethnic groups

Little Italy in New York, ca.1900.

Italy used to be a country of mass emigration from the late 19th century until the 1970s. Between 1898 and 1914, the peak years of Italian diaspora, approximately 750,000 Italians emigrated each year.[117] Italian communities once thrived in the former African colonies of Eritrea (nearly 100,000 at the beginning of World War II),[118] Somalia and Libya (150,000 Italians settled in Libya, constituting about 18% of the total population).[119] All of Libya's Italians were expelled from the North African country in 1970.[120] In addition, after the communist occupation of Istria in 1945, up to 350,000 ethnic Italians left Titoist Yugoslavia.[121] Today, large numbers of people with full or significant Italian ancestry are found in Brazil (25 million),[122] Argentina (20 million),[123] US (17.8 million),[124] France (5 million),[125] Uruguay (1.5 million),[126] Canada (1.4 million),[127] Venezuela (900,000)[128] and Australia (800,000).[129]

As a result of the profound economic and social changes induced by postwar industrialization, including low birth rates, an aging population and thus a shrinking workforce, during the 1980s Italy became to attract rising flows of foreign immigrants. The present-day figure of about 4.6 million foreign residents, that make up some 7.5% of the total population, include more than half a million children born in Italy to foreign nationals—second generation immigrants, but exclude foreign nationals who have subsequently acquired Italian nationality; this applied to 53,696 people in 2008.[130] The official figures also exclude illegal immigrants, the so-called *clandestini*, whose numbers are very difficult to determine. In May 2008 *The Boston Globe* quoted an estimate of 670,000 for this group.[131] Since the fall of the Berlin Wall in 1989, and more recently, the 2004 and 2007 enlargements of the European Union, the main waves of migration came from the former socialist countries of Eastern Europe (especially Romania, Albania, Ukraine and Poland). The second most important area of immigration to Italy has always been the neighbouring North Africa (in particular, Morocco, Egypt and Tunisia), with soaring arrivals as a consequence of the Arab Spring. Furthermore, in recent years, growing migration fluxes from the Far East (notably, China[132] and the Philippines) and Latin America (Ecuador, Peru) have been recorded. Currently, more than one million Romanians (around one tenth of them being Roma[133]) are officially registered as living in Italy, representing thus the most important individual country of origin, followed by Albanians and Moroccans with about 500,000 people each. The number of unregistered Romanians is difficult to estimate, but the Balkan Investigative Reporting Network suggested that in 2007 that there might have been half a million or more.[134] [135] Overall, at the end of 2000s (decade) the foreign born population of Italy was from: Europe (54%), Africa (22%), Asia (16%), the Americas (8%) and Oceania (0.06%). The distribution of immigrants is largely uneven in Italy: 87% of immigrants live in the northern and central parts of the country (the most economically developed areas), while only 13% live in the southern half of the peninsula.

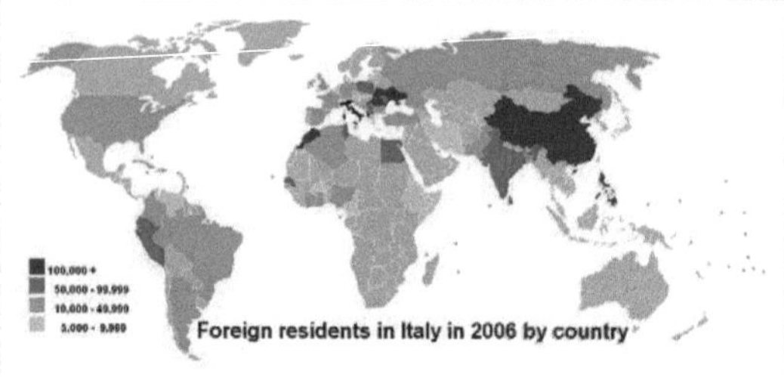

Estimated foreign-born population by country of birth, 2006 figures.

Origin of the population of Italy as of 1 January 2010

Origin	Population	Percent
Italian	56118099	92.47%
Romanian	1200000	1.98%
North African	646624	1.07%
Albanian	466684	0.77%
Chinese	188352	0.28%
Ukrainian	153998	0.31%
Asian (non-Chinese)	499013	0.83%
Latin American	324917	0.54%
Sub-Saharan African	285169	0.47%
Other	782549	1.29%

Languages

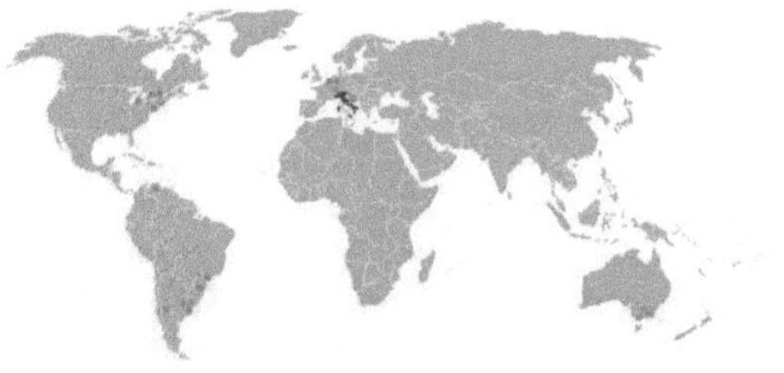

"Italophone" world. Official language. Secondary, widely spoken or understood. Understood by some due to former colonisation.

Italy's official language is Italian. Ethnologue has estimated that there are about 55 million speakers of the language in Italy and a further 6.7 million outside of the country.[136] However, between 120 and 150 million people use Italian as a second or cultural language, worldwide.[137]

Italian, adopted by the state after the unification of Italy, is based on the Florentine variety of Tuscan and is somewhat intermediate between the Italo-Dalmatian languages and the Gallo-Romance languages. Its development was also influenced by the Germanic languages of the post-Roman invaders.

Italy has numerous dialects spoken all over the country and some Italians cannot speak Italian at all.[138] However, the establishment of a national education system has led to decrease in variation in the languages spoken across the country. Standardisation was further expanded in the 1950s and 1960s thanks to economic growth and the rise of mass media and television (the state broadcaster RAI helped set a standard Italian).

Several linguistic groups are legally recognized,[139] and a number of minority languages have co-official status alongside Italian in various parts of the country. French is co-official in the Valle d'Aosta—although in fact Franco-Provencal is more commonly spoken there. German has the same status in the province of South Tyrol as, in some parts of that province and in parts of the neighbouring Trentino, does Ladin. Slovene is officially recognised in the provinces of Trieste, Gorizia and Udine in Friuli Venezia Giulia.

In these regions official documents are bilingual (trilingual in Ladin communities), or available upon request in either Italian or the co-official language. Traffic signs are also multilingual, except in the Valle d'Aosta where – with the exception of Aosta itself which has retained its Latin form in Italian (as in English) – French toponyms are generally used, attempts to italianise them during the Fascist period having been abandoned. Education is possible in minority languages where such schools are operating.

Religion

Religion in Italy, 2001[140]

Religion	Percent
Christianity	91.6%
None	5.8%
Islam	1.9%
Buddhism	0.3%
Hinduism	0.2%
Sikhism	0.1%
Judaism	0.1%

Roman Catholicism is by far the largest religion in the country, although the Catholic Church is no longer officially the state religion. The proportion of Italians that identify themselves as Roman Catholic is 87.8%,[141] although only about one-third of these described themselves as active members (36.8%). Most Italians believe in God, or a form of a spiritual life force. According to the most recent Eurobarometer Poll 2005:[142] 74% of Italian citizens responded that 'they believe there is a God', 16% answered that 'they believe there is some sort of spirit or life force' and 6% answered that 'they do not believe there is any sort of spirit, God, or life force'.

Christianity

The Italian Catholic Church is part of the global Roman Catholic Church, under the spiritual leadership of the Pope, curia in Rome, and the Conference of Italian Bishops. In addition to Italy, two other sovereign nations are included in Italian-based dioceses, the enclaves of San Marino and Vatican City. There are 225 dioceses in the Italian Catholic Church, see further in this article and in the article List of the Roman Catholic dioceses in Italy.

The Santa Maria del Fiore cathedral in Florence, which has the biggest brick dome in the world,[143] [144] and is considered a masterpiece of Italian architecture.

Italy has a rich Catholic culture, especially as numerous Catholic saints, martyrs and popes were Italian themselves. Roman Catholicism is the largest religion and denomination in Italy, with around 87.8% of Italians considering themselves Catholic. Italy is also home to the greatest number of cardinals in the world,[145] and is the country with the greatest number of Roman Catholic churches per capita.[146]

Even though the main Christian denomination in Italy is Roman Catholicism, there are relevant minorities of Protestant, Waldensians, Eastern Orthodox and other Christian churches. In the 20th century, Jehovah's Witnesses, Pentecostalism, non-denominational Evangelicalism, and Mormonism were the fastest-growing Protestant churches. Starting from the 1980s, Immigration from Subsaharan Africa has increased the size of Baptist, Anglican, Pentecostal and Evangelical communities in Italy, while immigration from Eastern Europe has established large Eastern Orthodox communities. At the beginning of 21st century, there were more than 700,000 Eastern Orthodox Christians in Italy, including 180,000 Greek Orthodox,[147] 550,000 Pentecostals and Evangelists (0.8%), of whom 400,000 are members of the Assemblies of God, 235,685 Jehovah's Witnesses (0.4%),[148] 30,000 Waldensians,[149] 25,000 Seventh-day Adventists, 22,000 Mormons, 15,000 Baptists (plus some 5,000 Free Baptists), 7,000 Lutherans, 4,000 Methodists (affiliated with the Waldensian Church).[150]

Other religions

The longest-established religious faith in Italy is Judaism, Jews having been present in Ancient Rome before the birth of Christ. Italy has seen many influential Italian-Jews, such as Shabbethai Donnolo (died 982), prime minister Luigi Luzzatti, who took office in 1910, and Ernesto Nathan, outstanding mayor of Rome from 1907 to 1913. During the Holocaust, Italy took in many Jewish refugees from Nazi Germany. However, with the creation of the Nazi-backed puppet Italian Social Republic, about 15% of Italy's Jews were killed, despite the Fascist government's refusal to deport Jews to Nazi death camps. This, together with the emigration that preceded and followed the Second World War, has left only a small community of around 45,000 Jews in Italy today. Due to rising immigration, there has been an increase in non-Christian faiths. In 2009, there were 1.0 million Muslims in Italy[151] forming 1.6 percent of population although, only 50,000 hold Italian citizenship. Independent estimates put the Islamic population in Italy anywhere from 0.8 million[152] to 1.5 million.[153] There are more than 200,000 followers of faiths originating in the Indian subcontinent with some 70,000 Sikhs with 22 gurdwaras across the country,[154] 70,000 Hindus, and 50,000 Buddhists.[155] There are an estimated some 4,900 Bahá'ís in Italy in 2005.[156]

Education

Italy's public education is free and compulsory from 6 to 15 years of age,[158] and has a five-year primary stage and an eight-year secondary stage, divided into first-grade secondary school (middle school) and second-grade secondary school (or high school). Italy has a high public education standard, surpassing that of other comparable developed countries, such as the UK and Germany. The country has both public and private education systems.

La Sapienza University in Rome is Europe's largest, with 140,000 students.[157]

According to National Science Indicators (1981–2002), a database produced by Research Services Group containing listings of output and citation statistics for more than 90 countries, Italy has an above-average output of scientific papers (in terms of number of papers written with at least one author being from Italy) in space science (9.75% of papers in the world being from Italy), mathematics (5.51% of papers in the world), computer science, neurosciences, and physics; the lowest, but still slightly above world-average, output in terms of number of papers produced is recorded in the social sciences, psychology and psychiatry, and economics and business.[159]

University of Bologna is Europe's oldest, founded in 1088.

Italy hosts a broad variety of universities, colleges and academies. Milan's Bocconi University, has been ranked among the top 20 best business schools in the world by The Wall Street Journal international rankings, especially thanks to its M.B.A. program, which in 2007 placed it no. 17 in the world in terms of graduate recruitment preference by major multinational companies.[160] Also, Forbes has ranked Bocconi no. 1 worldwide in the specific category Value for Money.[161] In May 2008, Bocconi overtook several traditionally top global business schools in the Financial Times Executive education ranking, reaching no. 5 in Europe and no. 15 in the world.[162]

Other top universities and polytechnics include the Polytechnic University of Turin, the Politecnico di Milano (which in 2009 was ranked as the 57th technical university in the world by Top Universities, in a research conducted on behalf of Times Higher Education.[163] This was a 6-positions growth from the 63rd position in 2008. In 2009 an Italian research ranked it as the best in Italy over indicators such as scientific production, attraction of foreign students, and others[164]), the University of Rome La Sapienza (which in 2005 was Europe's 33rd best university,[165] and ranks amongst Europe's 50 and the world's 150 best colleges[166]) and the University of Milan (whose research and teaching activities have developed over the years and have received important international recognitions. The University is the only Italian member of the League of European Research Universities (LERU), a prestigious group of twenty research-intensive European Universities. It also been awarded ranking positions as such: -1st in Italy and 7th in Europe (The Leiden Ranking – Universiteit Leiden).

Italy and the Western world's oldest college is the University of Bologna.[167] In 2009, the University of Bologna is, according to The Times, the only Italian college in the top 200 World Universities.

Health

Italian public health insurance card.

Italy has had a public healthcare system since 1978.[168] Healthcare spending in Italy accounted for more than 9.0% of the national GDP in 2008, slightly above the OECD countries' average of 8.9%.[169] However, Italy ranks as having the world's 2nd best healthcare system,[168] [170] and the world's 3rd best healthcare performance.[171]

Italy had the 12th highest worldwide life expectancy in 2010,[172] while, as in many others western countries, seeing an increase in the proportion of overweight and obese people, with 34.2% of Italians self reporting as overweight and 9.8% self reporting as obese.[173] The proportion of daily smokers was 22% in 2008.[174] Smoking in public places including bars, restaurants, night clubs and offices has been restricted to specially ventilated rooms since 2005.[175]

Culture

The city of Venice, built on 117 islands.

Italy did not exist as a state until the country's unification in 1861. Due to this comparatively late unification, and the historical autonomy of the regions that comprise the Italian Peninsula, many traditions and customs that are now recognized as distinctly Italian can be identified by their regions of origin. Despite the political and social distinction of these regions, Italy's contributions to the cultural and historical heritage of Europe and the world remain immense. Italy is home to the greatest number of UNESCO World Heritage Sites (47) to date, and has rich collections of art, culture and literature from many different periods. The country has had a broad cultural influence worldwide, also because numerous Italians emigrated to other places during the Italian diaspora. Furthermore, the nation has, overall, an estimated 100,000 monuments of any sort (museums, palaces, buildings, statues, churches, art galleries, villas, fountains, historic houses and archaeological remains).[176]

Architecture

The Leaning Tower of Pisa and the Duomo di Pisa, in the Piazza dei Miracoli. The Piazza is a UNESCO World Heritage Site,[177] and both buildings are examples of Romanesque architecture.[178]

Italy has a very broad and diverse architectural style, which cannot be simply classified by period, but also by region, due to Italy's division into several city-states until 1861. However, this has created a highly diverse and eclectic range in architectural designs. Italy is known for its considerable architectural achievements,[179] such as the construction of arches, domes and similar structures during ancient Rome, the founding of the Renaissance architectural movement in the late-14th to 16th century, and being the homeland of Palladianism, a style of construction which inspired movements such as that of Neoclassical architecture, and influenced the designs which noblemen built their country houses all over the world, notably in the UK, Australia and the US during the late-17th to early 20th centuries. Several of the finest works in Western architecture, such as the Colosseum, the Milan Cathedral and Florence cathedral, the Leaning Tower of Pisa and the building designs of Venice are found in Italy.

Italian architecture has also widely influenced the architecture of the world. British architect Inigo Jones, inspired by the designs of Italian buildings and cities, brought back the ideas of Italian Renaissance architecture to 17th century England, being inspired by Andrea Palladio.[180] Additionally, Italianate architecture, popular abroad since the 19th century, was used to describe foreign architecture which was built in an Italian style, especially modelled on Renaissance architecture.

Visual art

The Last Supper by Leonardo da Vinci: equally to *Mona Lisa*, it is the most famous, most reproduced and most parodied portrait and religious painting of all time.[181]

Over the centuries, Italian art has gone through many stylistic changes. Italian painting is traditionally characterized by a warmth of colour and light, as exemplified in the works of Caravaggio and Titian, and a preoccupation with religious figures and motifs. Italian painting enjoyed preeminence in Europe for hundreds of years, from the Romanesque and Gothic periods, and through the Renaissance and Baroque periods, the latter two of which saw fruition in Italy. Notable artists who fall within these periods include Michelangelo, Leonardo da Vinci, Donatello, Botticelli, Fra Angelico, Tintoretto, Caravaggio, Bernini, Titian and Raphael.

Thereafter, Italy was to experience a continual subjection to foreign powers which caused a shift of focus to political matters, leading to its decline as the artistic authority in Europe. Not until 20th century Futurism, primarily through the works of Umberto Boccioni and Giacomo Balla, would Italy recapture any of its former prestige as a seminal place of artistic evolution. Futurism was succeeded by the metaphysical paintings of Giorgio de Chirico, who exerted a strong influence on the Surrealists and generations of artists to follow.

Literature and theatre

Dante, poised between the mountain of purgatory and the city of Florence, displays the famous incipit *Nel mezzo del cammin di nostra vita* in a detail of Domenico di Michelino's painting, Florence 1465.

The basis of the modern Italian language was established by the Florentine poet Dante Alighieri, whose greatest work, the Divine Comedy, is considered amongst the foremost literary statements produced in Europe during the Middle Ages. There is no shortage of celebrated literary figures in Italy: Giovanni Boccaccio, Giacomo Leopardi, Alessandro Manzoni, Torquato Tasso, Ludovico Ariosto, and Petrarch, whose best-known vehicle of expression, the sonnet, was invented in Italy.

Prominent philosophers include Giordano Bruno, Marsilio Ficino, Niccolò Machiavelli, and Giambattista Vico. Modern literary figures and Nobel laureates are nationalist poet Giosuè Carducci in 1906, realist writer Grazia Deledda in 1926, modern theatre author Luigi Pirandello in 1936, poets Salvatore Quasimodo in 1959 and Eugenio Montale in 1975, satirist and theatre author Dario Fo in 1997.[182]

Italian theatre can be traced back to the Roman tradition which was heavily influenced by the Greek; as with many other literary genres, Roman dramatists tended to adapt and translate from the Greek. For example, Seneca's *Phaedra* was based on that of Euripides, and many of the comedies of Plautus were direct translations of works by Menander. During the 16th century and on into the 18th century, Commedia dell'arte was a form of improvisational theatre, and it is still performed today. Travelling troupes of players would set up an outdoor stage and provide amusement in the form of juggling, acrobatics, and, more typically, humorous plays based on a repertoire of established characters with a rough storyline, called *canovaccio*.

Music

Giacomo Puccini, Italian composer whose operas, including *La bohème*, *Tosca*, *Madama Butterfly*, and *Turandot*, are among the most frequently worldwide performed in the standard repertoire.[183] [184]

From folk music to classical, music has always played an important role in Italian culture. Instruments associated with classical music, including the piano and violin, were invented in Italy, and many of the prevailing classical music forms, such as the symphony, concerto, and sonata, can trace their roots back to innovations of 16th and 17th century Italian music.

Italy's most famous composers include the Renaissance composers Palestrina and Monteverdi, the Baroque composers Alessandro Scarlatti, Corelli and Vivaldi, the Classical composers Paganini and Rossini, and the Romantic composers Verdi and Puccini. Modern Italian composers such as Berio and Nono proved significant in the development of experimental and electronic music. While the classical music tradition still holds strong in Italy, as evidenced by the fame of its innumerable opera houses, such as La Scala of Milan and San Carlo of Naples, and performers such as the pianist Maurizio Pollini and the late tenor Luciano Pavarotti, Italians have been no less appreciative of their thriving contemporary music scene.

Italy is widely known for being the birthplace of opera.[185] Italian opera was believed to have been founded in the early 17th century, in Italian cities such as Mantua and Venice.[185] Later, works and pieces composed by native Italian composers of the 19th and early 20th centuries, such as Rossini, Bellini, Donizetti, Verdi and Puccini, are amongst the most famous operas ever written and today are performed in opera houses across the world. La Scala operahouse in Milan is also renowned as one of the best in the world. Famous Italian opera singers include Enrico Caruso, Alessandro Bonci, the late Luciano Pavarotti, and Andrea Bocelli, to name a few.

Luciano Pavarotti, one of the most famous tenors of all time.

Introduced in the early 1920s, jazz took a particularly strong foothold in Italy, and remained popular despite the xenophobic cultural policies of the Fascist regime. Today, the most notable centers of jazz music in Italy include Milan, Rome, and Sicily. Later, Italy was at the forefront of the progressive rock movement of the 1970s, with bands like PFM and Goblin. Italy was also an important country in the development of disco and electronic music, with Italo disco, known for its futuristic sound and prominent usage of synthesizers and drum machines, being one of the earliest electronic dance genres, as well as European forms of disco music aside from Euro disco (which later went on to influence several genres such as Eurodance and Nu-disco). Producers/songwriters such as Giorgio Moroder, who won three Academy Awards for his music, were highly influential in the development of EDM (electronic dance music). Today, Italian pop music is represented annually with the Sanremo Music Festival, which served as inspiration for the Eurovision song contest, and the Festival of Two Worlds in Spoleto. Singers such as pop diva Mina, classical crossover artist Andrea Bocelli, Grammy winner Laura Pausini, and European chart-topper Eros Ramazzotti have attained international acclaim.

Cinema

Federico Fellini, considered one of the most influential and widely revered filmmakers of the 20th century.[186]

The history of Italian cinema began a few months after the Lumière brothers began motion picture exhibitions. The first Italian film was a few seconds long, showing Pope Leo XIII giving a blessing to the camera. The Italian film industry was born between 1903 and 1908 with three companies: the Società Italiana Cines, the Ambrosio Film and the Itala Film. Other companies soon followed in Milan and in Naples. In a short time these first companies reached a fair producing quality, and films were soon sold outside Italy. Cinema was later used by Benito Mussolini, who founded Rome's renowned Cinecittà studio for the production of Fascist propaganda until World War II.[187]

After the war, Italian film was widely recognised and exported until an artistic decline around the 1980s. Notable Italian film directors from this period include Vittorio De Sica, Federico Fellini, Sergio Leone, Pier Paolo Pasolini, Luchino Visconti, Michelangelo Antonioni and Dario Argento. Movies include world cinema treasures such as *La dolce vita*, *The Good, the Bad and the Ugly* and *Bicycle Thieves*. The mid-1940s to the early 1950s was the heyday of neorealist films, reflecting the poor condition of post-war Italy.[188] [189] As the country grew wealthier in the 1950s, a form of neorealism known as pink neorealism succeeded, and other film genres, such as sword-and-sandal followed as spaghetti westerns, were popular in the 1960s and 1970s. In recent years, the Italian scene has received only occasional international attention, with movies like *La vita è bella* directed by Roberto Benigni and *Il postino* with Massimo Troisi.

Science

Galileo is considered one of the fathers of modern science.[190] [190] [191] [192]

Through the centuries, Italy has given birth to some notable scientific minds. Amongst them, and perhaps the most famous polymath in history, Leonardo da Vinci made several contributions to a variety of fields including art, biology, and technology. Galileo Galilei was a physicist, mathematician, and astronomer who played a major role in the Scientific Revolution. His achievements include improvements to the telescope and consequent astronomical observations, and support for Copernicanism. The physicist Enrico Fermi, a Nobel prize laureate, was the leader of the team that built the first nuclear reactor and is also noted for his many other contributions to physics, including the co-development of the quantum theory.

A brief overview of some other notable figures includes the astronomer Giovanni Domenico Cassini, who made many important discoveries about the Solar System; the physicist Alessandro Volta, inventor of the electric battery; the mathematicians Lagrange, Fibonacci, and Gerolamo Cardano, whose Ars Magna is generally recognized as the first modern treatment on mathematics, made fundamental advances to the field; Marcello Malpighi, a doctor and founder of microscopic anatomy; the biologist Lazzaro Spallanzani, who conducted important research in bodily functions, animal reproduction, and cellular theory; the physician, pathologist, scientist, and Nobel laureate Camillo Golgi, whose many achievements include the discovery of the Golgi complex, and his role in paving the way to the acceptance of the Neuron doctrine; and Guglielmo Marconi, who received the Nobel Prize in Physics for the invention of radio.

Sport

San Siro Stadium in Milan, with a capacity of over 80,000, is one of Europe's largest.

Italy has a long sporting tradition. In numerous sports, both individual and team, Italy has good representation and many successes. The most popular sport is by far football. Basketball and volleyball are the next most popular/played, with Italy having a rich tradition in both. Italy won the 2006 FIFA World Cup, and is currently the second most successful football team in the world, after Brazil, having won four FIFA World Cups.[193] Italy has also got strong traditions in cycling, tennis, athletics, fencing, winter sports and rugby. Italian Scuderia Ferrari is the oldest surviving team in Grand Prix racing, having competed since 1948, and statistically the most successful Formula One team in history with a record of 15 drivers' championships and 16 constructors' championships.

Fashion and design

A Milan Fashion Week event.

Italian fashion has a long tradition, and is regarded as one of the most important in the world. Milan, Florence and Rome are Italy's main fashion capitals. According to the 2009 Global Language Monitor, Milan was nominated the true fashion capital of the world, even surpassing other major capitals, such as New York, Paris, London and Tokyo, while Rome came 4th.[194] Major Italian fashion labels, such as Gucci, Prada, Versace, Valentino, Armani, Dolce & Gabbana, Missoni, Fendi, Moschino, Max Mara and Ferragamo, to name a few, are regarded as amongst the finest fashion houses in the world. Also, the fashion magazine Vogue Italia, is considered the most important and prestigious fashion magazine in the world.[195]

Italy is also prominent in the field of design, notably interior design, architectural design, industrial design and urban design. The country has produced some well-known furniture designers, such as Gio Ponti and Ettore Sottsass, and Italian phrases such as *"Bel Disegno"* and *"Linea Italiana"* have entered the vocabulary of furniture design.[196] Examples of classic pieces of Italian white goods and pieces of furniture include Zanussi's washing machines and fridges,[197] the "New Tone" sofas by Atrium,[197] and the post-modern bookcase by Ettore Sottsass, inspired by Bob Dylan's song "Stuck Inside of Mobile with the Memphis Blues Again".[197] Today, Milan and Turin are the nation's leaders in architectural design and industrial design. The city of Milan hosts the FieraMilano, Europe's biggest design fair.[198] Milan also hosts major design and architecture-related events and venues, such as the "*Fuori Salone*" and the Salone del Mobile, and has been home to the designers Bruno Munari, Lucio Fontana, Enrico Castellani and Piero Manzoni[199]

Cuisine

Original Neapolitan Pizza Margherita.

Modern Italian cuisine has evolved through centuries of social and political changes, with its roots reaching back to the 4th century BC. Significant change occurred with the discovery of the New World, when vegetables such as potatoes, tomatoes, bell peppers, and maize became available. However, these central ingredients of modern Italian cuisine were not introduced in scale before the 18th century.[200]

Ingredients and dishes vary by region. However, many dishes that were once regional have proliferated in different variations across the country. Cheese and wine are major parts of the cuisine, playing different roles both regionally and nationally with their many variations and *Denominazione di origine controllata* (regulated appellation) laws. Coffee, and more specifically espresso, has become highly important to the cultural cuisine of Italy. Some famous dishes and items include pasta, pizza, *lasagna*, *focaccia*, and *gelato*.

See also

- Outline of Italy
- Index of Italy-related articles
- List of English exonyms for Italian toponyms

Notes

[1] "Ethnologue report" (http://www.ethnologue.com/show_country.asp?name=IT). Ethnologue.com. . Retrieved 30 October 2010.

[2] "Monthly demographic balance: January 2011" (http://demo.istat.it/bilmens2011gen/index.html) (in Italian). Istat. 10 September 2011. . Retrieved 10 September 2011.

[3] "Italy" (http://www.imf.org/external/pubs/ft/weo/2012/01/weodata/weorept.aspx?pr.x=34&pr.y=12&sy=2009&ey=2012&scsm=1&ssd=1&sort=country&ds=.&br=1&c=136&s=NGDPD,NGDPDPC,PPPGDP,PPPPC,LP&grp=0&a=). International Monetary Fund. . Retrieved 18 April 2012.

[4] "Distribution of family income – Gini index" (https://www.cia.gov/library/publications/the-world-factbook/fields/2172.html). CIA – The World Factbook. . Retrieved 9 July 2010.

[5] "Human Development Report 2011" (http://hdr.undp.org/en/media/HDR_2011_EN_Table1.pdf). United Nations. 2011. . Retrieved 5 November 2011.

[6] "Comune di Campione d'Italia" (http://www.comune.campione-d-italia.co.it/). Comune.campione-d-italia.co.it. 14 July 2010. . Retrieved 30 October 2010.

[7] The English name of the Italian state used in the English text of the Treaty of Lisbon,

[8] As used in the following UN memorandum: "UNTERM database of the United Nations" (http://unterm.un.org/dgaacs/unterm.nsf/c2f5f1ea9d52ce7485256dc5006e5940/a7c582eab561fa3f852573b600767fad?OpenDocument). United Nations Secretariat. . Retrieved 28 March 2012.

[9] In Italy, other languages have been officially recognised as legitimate autochthonous (regional) languages under the European Charter for Regional or Minority Languages. In each of these, Italy's official name is as follows:

- Emilian: *Ripòbblica Itaglièna*;
- Friulian: *Republiche Taliane*;
- Ladin: *Republica taliana*;
- Lombard: *Repùblega Taliana*;
- Neapolitan: *Repubbreca italiana*;
- Piemontese: *Repùblica Italian-a*;
- Sardinian: *Repubrica Italiana*;
- Sicilian: *Ripùbblica di Tàlia*;
- Venetian: *Republica Tałiana*

[10] "Italy: Birthplace of the Renaisssance" (http://www.immaculateheartacademy.org/outside2/socialstudies/kuhns/1 Italy Birthplace of the Renaissance.pdf). *European Renaissance and Reformation*. Township of Washington, NJ: Immaculate Heart Academy. [n.d.]. . Retrieved 20 December 2009.

[11] "Unification of Italy" (http://library.thinkquest.org/TQ0312582/unification.html). Library.thinkquest.org. 4 April 2003. . Retrieved 19 November 2009.
[12] "The Italian Colonial Empire" (http://www.allempires.com/article/index.php?q=italian_colonial). All Empires. . Retrieved 30 October 2010.
[13] The Economist Intelligence Unit's quality-of-life index (http://www.economist.com/media/pdf/QUALITY_OF_LIFE.pdf), Economist, 2005
[14] "Report for Selected Countries and Subjects" (http://imf.org/external/pubs/ft/weo/2009/02/weodata/weorept.aspx?pr.x=26&pr.y=9&sy=2009&ey=2009&scsm=1&ssd=1&sort=country&ds=.&br=1&c=512,941,914,446,612,666,614,668,311,672,213,946,911,137,193,962,122,674,912,676,313,548,419,556,513,678,316,181,913,682,124,684,339,273,638 s=NGDPDPC&grp=0&a=). Imf.org. 14 September 2006. . Retrieved 2 August 2010.
[15] "DDP Quick Query" (http://ddp-ext.worldbank.org/ext/DDPQQ/member.do?method=getMembers&userid=1&queryId=135). Ddp-ext.worldbank.org. 20 July 2004. . Retrieved 2 August 2010.
[16] "Report for Selected Countries and Subjects" (http://www.imf.org/external/pubs/ft/weo/2009/01/weodata/weorept.aspx?sy=2008&ey=2008&scsm=1&ssd=1&sort=country&ds=.&br=1&c=512,941,914,446,612,666,614,668,311,672,213,946,911,137,193,962,122,674,912,676,313,548,419,556,513,678,316,181,913,682,124,684,339,273,638 s=PPPGDP&grp=0&a=&pr.x=9&pr.y=12). Imf.org. 14 September 2006. . Retrieved 2 August 2010.
[17] "CIA World Factbook, Budget" (https://www.cia.gov/library/publications/the-world-factbook/fields/2056.html). Cia.gov. . Retrieved 26 January 2011.
[18] M. De Leonardis, *Il Mediterraneo nella politica estera italiana del secondo dopoguerra*, Bologna, Il Mulino, 2003, p. 17
[19] *The Middle East and Europe* (http://books.google.com/?id=X4xw8-Oj9usC&pg=PA157&lpg=PA157&dq=regional+power+italy+in+europe#PPP1,M1). Google Books. 1998-11-24. . Retrieved 2011-05-30.
[20] "KOF – Pressemitteilung" (http://globalization.kof.ethz.ch/static/pdf/press_release_2009_en.pdf) (PDF). . Retrieved 27 October 2009.
[21] Alberto Manco, *Italia. Disegno storico-linguistico*, 2009, Napoli, L'Orientale, ISBN 978-88-95044-62-0
[22] OLD, p. 974: "first syll. naturally short (cf. Quint.*Inst.*1.5.18), and so scanned in Lucil.825, but in dactylic verse lengthened *metri gratia*."
[23] J.P. Mallory and D.Q. Adams, *Encyclopedia of Indo-European Culture* (London: Fitzroy and Dearborn, 1997), 24.
[24] Dionysius of Halicarnassus, *Roman Antiquities*, 1.35 (http://penelope.uchicago.edu/Thayer/E/Roman/Texts/Dionysius_of_Halicarnassus/1B*.html), on LacusCurtius
[25] Aristotle, *Politics*, 7.1329b (http://www.perseus.tufts.edu/hopper/text?doc=Perseus:text:1999.01.0058:book=7:section=1329b#note-link2), on Perseus
[26] Thucydides, *The Peloponnesian War*, 6.2.4 (http://www.perseus.tufts.edu/hopper/text?doc=Thuc.+6.2.4&fromdoc=Perseus:text:1999.01.0200), on Perseus
[27] Pallottino, M., History of Earliest Italy, trans. Ryle, M & Soper, K. in Jerome Lectures, Seventeenth Series, p. 50
[28] Kluwer Academic/Plenum Publishers 2001, ch. 2. ISBN 0-306-46463-2.
[29] The [[Mycenaeans (http://www.gla.ac.uk/departments/archaeology/research/projects/mycenaeansitaly/)] and Italy: the archaeological and archaeometric ceramic evidence], University of Glasgow, Department of Archaeology
[30] Emilio Peruzzi, *Mycenaeans in early Latium*, (Incunabula Graeca 75), Edizioni dell'Ateneo & Bizzarri, Roma, 1980
[31] Lord William Taylour, Mycenaean Pottery in Italy and Adjacent Areas (Cambridge 1958)
[32] Gert Jan van Wijngaarden, *Use and Appreciation of Mycenaean Pottery in the Levant, Cyprus and Italy (1600-1200 B.C.): The Significance of Context*, Amsterdam Archaeological Studies, Amsterdam University Press, 2001
[33] Andrea Vianello, *Late Bronze Age Mycenaean and Italic Products in the West Mediterranean: A Social and Economic Analysis*, (British Archaeological Reports International Series), British Archaeological Reports
[34] Miriam S. Balmuth, Robert J. Rowland, *Studies in Sardinian archaeology*, University of Michigan Press, 1984
[35] Bryan Feuer, *Mycenaean civilization: an annotated bibliography through 2002*, McFarland & Company; Rev Sub edition (2 March 2004)
[36] Stéphane Barry and Norbert Gualde, "The Biggest Epidemics of History" (La plus grande épidémie de l'histoire), in L'Histoire n° 310, June 2006, pp. 45–46
[37] " Plague (http://www.brown.edu/Departments/Italian_Studies/dweb/plague/effects/death_toll.shtml)". Brown University.
[38] Jensen 1992, p. 64.
[39] Karl Julius Beloch, *Bevölkerungsgeschichte Italiens*, volume 3, pp. 359–360.
[40] Thomas James Dandelet, John A. Marino (2007). *Spain in Italy: politics, society, and religion 1500-1700*. Leiden: Koninklijke Brill. ISBN 978-90-04-15429-2.
[41] Napoleon Bonaparte, "The Economy of the Empire in Italy: Instructions from Napoleon to Eugène, Viceroy of Italy," *Exploring the European Past: Texts & Images*, Second Edition, ed. Timothy E. Gregory (Mason: Thomson, 2007), 65-66.
[42] (Mack Smith, Denis (1997). *Modern Italy; A Political History*. Ann Arbor: The University of Michigan Press. ISBN 0-472-10895-6
[43] (Bosworth (2005), pp. 49.)
[44] Mortara, G (1925). *La Salute pubblica in Italia durante e dopo la Guerra*. New Haven: Yale University Press.
[45] "Italy – Britannica Online Encyclopedia" (http://www.britannica.com/EBchecked/topic/297474/Italy#). Britannica.com. . Retrieved 2 August 2010.
[46] Adrian Lyttelton (editor), *"Liberal and fascist Italy, 1900–1945"*, Oxford University Press, 2002. pp. 13

[47] *Damage Foreshadows A-Bomb Test , 1946/06/06 (1946)* (http://www.archive.org/details/1946-06-06_Damage_Foreshadows_A-Bomb_Test). Universal Newsreel. 1946. . Retrieved February 22, 2012.

[48] "Italia 1946: le donne al voto, dossier a cura di Mariachiara Fugazza e Silvia Cassamagnaghi" (http://www.insmli.it/pubblicazioni/35/Voto donne versione def.pdf) (PDF). . Retrieved 2011-05-30.

[49] **(Italian)** "Commissione parlamentare d'inchiesta sul terrorismo in Italia e sulle cause della mancata individuazione dei responsabili delle stragi (Parliamentary investigative commission on terrorism in Italy and the failure to identify the perpetrators)" (http://web.archive.org/web/20060819211212/http://www.isn.ethz.ch/php/documents/collection_gladio/report_ital_senate.pdf). 1995. Archived from the original (http://www.isn.ethz.ch/php/documents/collection_gladio/report_ital_senate.pdf) on 2006-08-19. . Retrieved 2006-05-02.

[50] **(English)/(Italian)/(French)/(German)** "Secret Warfare: Operation Gladio and NATO's Stay-Behind Armies" (http://web.archive.org/web/20060425182721/http://www.isn.ethz.ch/php/collections/coll_gladio.htm#Documents). Swiss Federal Institute of Technology / International Relation and Security Network. Archived from the original (http://www.isn.ethz.ch/php/collections/coll_gladio.htm#Documents) on 2006-04-25. . Retrieved 2006-05-02.

[51] "Clarion: Philip Willan, Guardian, 24 June 2000, page 19" (http://www.cambridgeclarion.org/press_cuttings/us.terrorism_graun_24jun2000.html). Cambridgeclarion.org. 2000-06-24. . Retrieved 2010-04-24.

[52] Official French maps show the border detouring south of the main summit, and claim the highest point in Italy is Mont Blanc de Courmayeur (4,748 m), but these are inconsistent with an 1861 convention and topographic watershed analysis.

[53] "Morphometric and hydrological characteristics of some important Italian lakes" (http://www.iii.to.cnr.it/limnol/cicloac/lagit.htm). Largo Tonolli 50, 28922 Verbania Pallanza: Istituto per lo Studio degli Ecosistemi. . Retrieved 3 March 2010.

[54] "Italy – Environment" (http://dev.prenhall.com/divisions/hss/worldreference/IT/environment.html). Dev.prenhall.com. . Retrieved 2 August 2010.

[55] "National Parks in Italy" (http://www.parks.it/indice/NatParks.html). Parks.it. 1995-2010. . Retrieved 15 March 2010.

[56] REN21 (15 July 2010). "Renewables 2010 Global Status Report" (http://www.ren21.net/Portals/97/documents/GSR/REN21_GSR2011.pdf). REN21. . Retrieved 16 July 2010.

[57] "Photovoltaic energy barometer 2010 – EurObserv'ER" (http://www.eurobserv-er.org/pdf/baro196.asp). . Retrieved 30 October 2010.

[58] "World Wind Energy Report 2010" (http://www.wwindea.org/home/images/stories/pdfs/worldwindenergyreport2010_s.pdf) (PDF). *Report*. World Wind Energy Association. February 2011. . Retrieved 8-August-2011.

[59] wwea

[60] "Italy – Environment" (http://www.nationsencyclopedia.com/Europe/Italy-ENVIRONMENT.html). Encyclopedia of the Nations. . Retrieved 7 April 2010.

[61] United Nations Statistics Division, Millennium Development Goals indicators: Carbon dioxide emissions (CO_2), thousand metric tons of CO_2 (http://mdgs.un.org/unsd/mdg/SeriesDetail.aspx?srid=749&crid=) (collected by CDIAC)

[62] Human-produced, direct emissions of carbon dioxide only. Excludes other greenhouse gases; land-use, land-use-change and forestry (LULUCF); and natural background flows of CO_2 (See also: Carbon cycle)

[63] (http://www.euro.who.int/document/hms/ehiexes_e.pdf)

[64] Italy nuclear: Berlusconi accepts referendum blow, BBC News Europe, 14 June 2011 (http://www.bbc.co.uk/news/world-europe-13741105)

[65] Nick Squires (2 October 2009). "Sicily mudslide leaves scores dead" (http://www.telegraph.co.uk/news/worldnews/europe/italy/6255575/Sicily-mudslide-leaves-scores-dead.html#). London: The Daily Telegraph. . Retrieved 2 October 2009.

[66] Adriana Rigutti, Meteorologia, Giunti, p, 95, 2009.

[67] Thomas A. Blair, Climatology: General and Regional, Prentice Hall pages 131–132

[68] Regions in *Italics* have an autonomous status.

[69] Smyth, Howard McGaw Italy: From Fascism to the Republic (1943–1946) (http://links.jstor.org/sici?sici=0043-4078(194809)1:3<205:IFFTTR>2.0.CO;2-O) *The Western Political Quarterly* vol. 1 no. 3 (pp. 205–222), September 1948

[70] Claudio Tucci (11 November 2008). "Confesercenti, la crisi economica rende ancor più pericolosa la mafia" (http://www.ilsole24ore.com/art/SoleOnLine4/Economia e Lavoro/2008/11/confesercenti-mafia-racket-pizzo.shtml?uuid=20ff3b9c-afe7-11dd-8057-9c09c8bfa449) (in Italian). *Confesercenti*. Ilsole24ore.com. . Retrieved 21 April 2011.

[71] Nick Squires (09 Jan 2010). "Italy claims finally defeating the mafia" (http://www.telegraph.co.uk/news/worldnews/europe/italy/6957240/Italy-claims-finally-defeating-the-mafia.html). Telegraph.co.uk. . Retrieved 21 April 2011.

[72] Kiefer, Peter (October 22, 2007). "Mafia crime is 7% of GDP in Italy, group reports" (http://www.nytimes.com/2007/10/22/world/europe/22iht-italy.4.8001812.html?_r=1). *The New York Times*. . Retrieved April 19, 2011.

[73] Maria Loi (October 1, 2009). "Rapporto Censis: 13 milioni di italiani convivono con la mafia" (http://www.antimafiaduemila.com/content/view/20052/78/) (in Italian). *Censis*. Antimafia Duemila. . Retrieved April 21, 2011.

[74] Kington, Tom (1 October 2009). "Mafia's influence hovers over 13m Italians, says report" (http://www.guardian.co.uk/world/2009/oct/01/mafia-influence-hovers-over-italians). *The Guardian* (London). . Retrieved 5 May 2010.

[75] ANSA (March 14, 2011). "Italy: Anti-mafia police arrest 35 suspects in northern Lombardy region" (http://mafiatoday.com/sicilian-mafia-ndrangheta/italy-anti-mafia-police-arrest-35-suspects-in-northern-lombardy-region/). *adnkronos.com*. Mafia Today. . Retrieved April 21, 2011.

[76] "Crime Statistics > Murders (per capita) (most recent) by country" (http://www.nationmaster.com/graph/cri_mur_percap-crime-murders-per-capita). NationMaster.com. . Retrieved 4 April 2010.

[77] " Italian soldiers leave for Lebanon (http://www.corriere.it/Primo_Piano/Cronache/2006/08_Agosto/29/libano.shtml) Corriere della Sera, 30 August 2006

[78] "Law n°6433 of September, 3 1999" (http://www.difesa.it/Difesa-cittadino/Riforma+del+Servizio+Militare.htm). Difesa.it. . Retrieved 30 October 2010.

[79] "The Military Balance 2010", pp. 141–145. International Institute for Strategic Studies, 3 February 2010.

[80] Italian Ministry of Defence. "Nota aggiuntiva allo stato di previsione per la Difesa per l'anno 2009" (http://www.difesa.it/NR/rdonlyres/5EF11493-59DD-4FB7-8485-F4258D9F5891/0/Nota_Aggiuntiva_2009.pdf) (in Italian). . Retrieved 27 April 2009.

[81] Hans M. Kristensen / Natural Resources Defense Council (2005). "NRDC: U.S. Nuclear Weapons in Europe – part 1" (http://www.nrdc.org/nuclear/euro/euro_pt1.pdf) (PDF). . Retrieved 2011-05-30.

[82] "Marina Militare (Italian military navy website)" (http://www.marina.difesa.it/) (in Italian). Marina.difesa.it. . Retrieved 2011-05-30.

[83] "The Carabinieri Force is linked to the Ministry of Defence" (http://www.carabinieri.it/Internet/Multilingua/EN/GoverningBodies/). Carabinieri. . Retrieved 14 May 2010.

[84] http://siteresources.worldbank.org/DATASTATISTICS/Resources/GDP.pdf

[85] http://siteresources.worldbank.org/DATASTATISTICS/Resources/GDP_PPP.pdf

[86] "Hobsons Careers in Europe | Country profiles" (http://www.careersineurope.hobsons.com/country_italy.aspx). Careersineurope.hobsons.com. . Retrieved 2 August 2010.

[87] "GDP per capita in PPS" (http://epp.eurostat.ec.europa.eu/cache/ITY_PUBLIC/2-25062009-BP/EN/2-25062009-BP-EN.PDF). Eurostat. . Retrieved 25 June 2009.

[88] (http://tg24.sky.it/tg24/economia/2010/06/14/ocse_tasso_disoccupazione_invariato_1.html) Ocse, tasso di disoccupazione stabile nell'eurozona

[89] "Italy – Economics" (http://dev.prenhall.com/divisions/hss/worldreference/IT/economics.html). Dev.prenhall.com. . Retrieved 2 August 2010.

[90] Pisa, Nick (12 June 2011). "Italy overtakes France to become world's largest wine producer" (http://www.telegraph.co.uk/foodanddrink/wine/8571222/Italy-overtakes-France-to-become-worlds-largest-wine-producer.html). *The Telegraph*. . Retrieved 17 August 2011.

[91] "Knowledge Economy Forum 2008: Innovative Small And Medium Enterprises Are Key To Europe & Central Asian Growth" (http://web.worldbank.org/WBSITE/EXTERNAL/COUNTRIES/ECAEXT/0,,contentMDK:21808326~menuPK:258604~pagePK:2865106~piPK:2865128~theSitePK:258599,00.html). The World Bank. 19 May 2005. . Retrieved 17 June 2008.

[92] "2010 Press Releases – Trade to expand by 9.5% in 2010 after a dismal 2009, WTO reports – Press/598" (http://www.wto.org/english/news_e/pres10_e/pr598_e.htm). WTO. . Retrieved 2011-05-30.

[93] "CIA – The World Factbook" (https://www.cia.gov/library/publications/the-world-factbook/geos/it.html). CIA. . Retrieved 26 January 2011.

[94] http://mkt.unwto.org/sites/all/files/docpdf/unwtohighlights11enhr_1.pdf

[95] Nicholas Crafts, Gianni Toniolo (1996). *Economic growth in Europe since 1945*. Cambridge University Press. p. 428. ISBN 0-521-49627-6.

[96] Eurostat. "Real GDP growth rate – Growth rate of GDP volume – percentage change on previous year" (http://epp.eurostat.ec.europa.eu/tgm/table.do?tab=table&init=1&plugin=1&language=en&pcode=tsieb020). . Retrieved 10 May 2009.

[97] "Europolitics" (http://www.europolitics.info/economy-monetary-affairs/deficits-increase-in-eurozone-and-eu-artb287086-50.html). Europolitics.info. . Retrieved 26 January 2011.

[98] "Could Italy Be Better Off than its Peers?" (http://www.cnbc.com/id/37207942/Could_Italy_Be_Better_Off_than_its_Peers). CNBC. 2010-05-18. . Retrieved 2011-05-30.

[99] "EUROPA – Press Releases – Regional GDP per inhabitant in the EU27, GDP per inhabitant in 2006 ranged from 25% of the EU27 average in Nord-Est in Romania to 336% in Inner London" (http://europa.eu/rapid/pressReleasesAction.do?reference=STAT/09/23&format=HTML&aged=0&language=EN&guiLanguage=en). Europa. 19 February 2009. . Retrieved 30 October 2010.

[100] "The real sick man of Europe" (http://www.economist.com/opinion/displaystory.cfm?story_id=3987219). The Economist. 19 May 2005. . Retrieved 10 May 2009.

[101] "Italy: The sick man of Europe" (http://www.telegraph.co.uk/comment/3557277/Italy-The-sick-man-of-Europe.html). London: The Daily Telegraph. 29 December 2008. . Retrieved 10 May 2009.

[102] Eurostat. "Energy, transport and environment indicators" (http://epp.eurostat.ec.europa.eu/cache/ITY_OFFPUB/KS-DK-08-001/EN/KS-DK-08-001-EN.PDF). . Retrieved 10 May 2009.

[103] Eurostat. "Panorama of energy" (http://epp.eurostat.ec.europa.eu/cache/ITY_OFFPUB/KS-GH-09-001/EN/KS-GH-09-001-EN.PDF). . Retrieved 10 May 2009.

[104] "European Cohesion Policy in Italy" (http://ec.europa.eu/regional_policy/sources/docgener/informat/country2009/it_en.pdf) (PDF). . Retrieved 30 October 2010.

[105] "Index of Economic Freedom" (http://web.archive.org/web/20080503060552/http://www.heritage.org/research/features/index/country.cfm?ID=Italy). The Heritage Foundation. Archived from the original (http://www.heritage.org/research/features/index/country.cfm?ID=Italy) on 3 May 2008. . Retrieved 4 November 2008.

[106] Eurostat. "R&D Expenditure and Personnel" (http://epp.eurostat.ec.europa.eu/cache/ITY_OFFPUB/KS-SF-08-091/EN/KS-SF-08-091-EN.PDF). . Retrieved 10 May 2009.

[107] Kiefer, Peter (22 October 2007). "Mafia crime is 7% of GDP in Italy, group reports" (http://www.nytimes.com/2007/10/22/world/europe/22iht-italy.4.8001812.html). *New York Times*. .

[108] "Nuovo record di velocità sulla linea AV Milano-Bologna" (http://209.85.129.132/search?q=cache:m5-u7afI-asJ:www.fsnews.it/fsnews/v/index.jsp?vgnextoid=07a4a4b1fb478110VgnVCM1000001c42fe0aRCRD) (in Italian). FSNews. 3 March 2008. Archived from the original (http://www.fsnews.it/fsnews/v/index.jsp?vgnextoid=07a4a4b1fb478110VgnVCM1000001c42fe0aRCRD) on 22 August 2008. . Retrieved 31 December 2008.

[109] "Record italiano di velocità" (http://www.duegieditrice.it/tuttotreno/default.asp?num=218) (in Italian). Tutto Treno. April 2008. . Retrieved 2 January 2009.

[110] "200 giorni al primo treno Alta Velocità sulla Milano-Bologna" (http://www.fsnews.it/cms/v/index.jsp?vgnextoid=d203c7bd6d20b110VgnVCM1000003f16f90aRCRD) (in Italian). FSNews. 28 May 2008. . Retrieved 31 December 2008.

[111] European Commission. "Panorama of Transport" (http://epp.eurostat.ec.europa.eu/cache/ITY_OFFPUB/KS-DA-07-001/EN/KS-DA-07-001-EN.PDF) (PDF). . Retrieved 3 May 2009.

[112] Istituto Nazionale di Statistica. "Resident population on 1st January" (http://dati.istat.it/Index.aspx?lang=en). . Retrieved 2011-12-19.

[113] >Istituto Nazionale di Statistica. "Foreign resident population on 1st January" (http://dati.istat.it/Index.aspx?lang=en). . Retrieved 2011-12-19.

[114] EUROSTAT. "Ageing characterises the demographic perspectives of the European societies - Issue number 72/2008" (http://epp.eurostat.ec.europa.eu/cache/ITY_OFFPUB/KS-SF-08-072/EN/KS-SF-08-072-EN.PDF). . Retrieved 28 April 2009.

[115] **(Italian)** ISTAT. "Crude birth rates, mortality rates and marriage rates 2005-2008" (http://demo.istat.it/altridati/indicatori/2008/Tab_1.pdf). . Retrieved 10 May 2009.

[116] **(Italian)** ISTAT. "Average number of children born per woman 2005-2008" (http://demo.istat.it/altridati/indicatori/2008/Tab_4.pdf). . Retrieved 3 May 2009.

[117] "Causes of the Italian mass emigration" (http://library.thinkquest.org/26786/en/articles/view.php3?arKey=4&paKey=7&loKey=0&evKey=&toKey=&torKey=&tolKey=). ThinkQuest Library. 15 August 1999. . Retrieved 30 October 2010.

[118] "Essay on Italian emigration to Eritrea (in Italian)" (http://www.ilcornodafrica.it/rds-01emigrazione.pdf) (PDF). . Retrieved 30 October 2010.

[119] Libya – Italian colonization (http://www.britannica.com/EBchecked/topic/339574/Libya/46562/Italian-colonization). Britannica Online Encyclopedia.

[120] Libya cuts ties to mark Italy era. (http://news.bbc.co.uk/2/hi/africa/4380360.stm). BBC News. 27 October 2005.

[121] Election Opens Old Wounds In Trieste (http://www.nytimes.com/1987/06/06/world/election-opens-old-wounds-in-trieste.html). The New York Times. 6 June 1987.

[122] Consulta Nazionale Emigrazione. Progetto ITENETs – "Gli italiani in Brasile"; pp. 11, 19 (http://www.consultanazionaleemigrazione.it/itestero/Gli_italiani_in_Brasile.pdf) . Retrieved 10 September 2008.

[123] **(Spanish)** Lee, Adam (3 April 2006). "Unos 20 millones de personas que viven en la Argentina tienen algún grado de descendencia italiana" (http://www.asteriscos.tv/dossier-3.html) (in Spanish). . Retrieved 27 June 2008.

[124] American FactFinder, United States Census Bureau. "U.S Census Bureau – Selected Population Profile in the United States" (http://factfinder.census.gov/servlet/IPTable?_bm=y&-reg=ACS_2006_EST_G00_S0201:543;ACS_2006_EST_G00_S0201PR:543;ACS_2006_EST_G00_S0201T:543;ACS_2006_EST_G00_S0201TPR:543&-qr_name=ACS_2006_EST_G00_S0201&-qr_name=ACS_2006_EST_G00_S0201PR&-qr_name=ACS_2006_EST_G00_S0201T&-qr_name=ACS_2006_EST_G00_S0201TPR&-ds_name=ACS_2006_EST_G00_&-TABLE_NAMEX=&-ci_type=A&-redoLog=true&-charIterations=047&-geo_id=01000US&-geo_id=NBSP&-format=&-_lang=en). American FactFinder, United States Census Bureau. . Retrieved 2011-05-30.

[125] " *The Cambridge survey of world migration* (http://books.google.com/books?id=BLo2RqGdv_wC&pg=PA143&dq&hl=en#v=onepage&q=&f=false)". Robin Cohen (1995). Cambridge University Press. p. 143. ISBN 0-521-44405-5

[126] "Ethnic origins, 2006 counts, for Uruguay, provinces and territories – 20% sample data" (http://www.hotelsclick.com/hoteles/UY/Uruguay-DEMOGRAFÃ-A-5.html). .

[127] "Ethnic origins, 2006 counts, for Canada, provinces and territories – 20% sample data" (http://www12.statcan.ca/english/census06/data/highlights/ethnic/pages/Page.cfm?Lang=E&Geo=PR&Code=01&Data=Count&Table=2&StartRec=1&Sort=3&Display=All&CSDFilter=5000). .

[128] Santander Laya-Garrido, Alfonso. *Los Italianos forjadores de la nacionalidad y del desarrollo economico en Venezuela*. Editorial Vadell. Valencia, 1978

[129] "20680-Ancestry by Country of Birth of Parents – Time Series Statistics (2001, 2006 Census Years) – Australia" (http://www.censusdata.abs.gov.au/ABSNavigation/prenav/ViewData?action=404&documentproductno=0&documenttype=Details&order=1&tabname=Details&areacode=0&issue=2006&producttype=Census Tables&javascript=true&textversion=false&navmapdisplayed=true&breadcrumb=LPTD&&collection=Census&period=2006&productlabel=Ancestry by Country of Birth of Parents - Time Series Statistics (2001, 2006 Census Years)&producttype=Census Tables&method=Place of Usual Residence&topic=Ancestry&). Australian Bureau of Statistics. 27 June 2007. . Retrieved 30 December 2008.

[130] "La popolazione straniera residente in Italia al 1° gennaio 2009 [The Foreign Population Resident in Italy on 1 January 2009]" (http://www.istat.it/salastampa/comunicati/non_calendario/20091008_00/testointegrale20091008.pdf) (in Italian) (PDF). Istat. 8 October 2009. pp. 1–3. . Retrieved 27 October 2009.

[131] Elisabeth Rosenthal, " Italy cracks down on illegal immigration (http://www.boston.com/news/world/europe/articles/2008/05/16/italy_cracks_down_on_illegal_immigration/)". The Boston Globe. 16 May 2008.
[132] " Milan police in Chinatown clash (http://news.bbc.co.uk/2/hi/europe/6550725.stm)". BBC News. 13 April 2007.
[133] " EUROPE: Home to Roma, And No Place for Them (http://www.ipsnews.net/news.asp?idnews=42404)". IPS ipsnews.net.
[134] "Balkan Investigative Reporting Network" (http://www.birn.eu.com/en/111/15/5745/). Birn.eu.com. 08 11 2007. . Retrieved 4 November 2008.
[135] According to Mitrica, an October 2005 Romanian report estimates that 1,061,400 Romanians are living in Italy, constituting 37% of 2.8 million immigrants in that countryMitrica, Mihai Un milion de romani s-au mutat in Italia (http://web.archive.org/web/20071023072029/http://evz.ro/article.php?artid=201813) ("One million Romanians have moved to Italy"). *Evenimentul Zilei*, 31 October 2005. Visited 11 April 2006. but it is unclear how the estimate was made, and therefore whether it should be taken seriously.
[136] Ethnologue report for language code:ita (Italy) (http://www.ethnologue.com/show_language.asp?code=ita) – Gordon, Raymond G., Jr. (ed.), 2005. Ethnologue: Languages of the World, Fifteenth edition. Dallas, Tex.: SIL International. Online version
[137] Italian Language: Geographic Distribution (http://www.servinghistory.com/topics/Italian_language::sub::Geographic_Distribution) *Discovery Media*. Retrieved 16 May 2010.
[138] "Italian language – Britannica Online Encyclopedia" (http://www.britannica.com/EBchecked/topic/297241/Italian-language). Britannica.com. 3 November 2008. . Retrieved 19 November 2009.
[139] [L.cost. 26 febbraio 1948, n. 4, Statuto speciale per la Valle d'Aosta; L.cost. 26 febbraio 1948, n. 5, Statuto speciale per il Trentino-Alto Adige; L.cost. 31 gennaio 1963, n. 1, Statuto speciale della Regione Friuli Venezia Giulia]
[140] Religious Populations (http://www.statistics.gov.uk/cci/nugget.asp?id=954) Office for National Statistics
[141] "Italy: 88% of Italians declare themselves Catholic" (http://www.corriere.it/Primo_Piano/Cronache/2006/01_Gennaio/17/cattolici.shtml) (in Italian). Corriere della Sera. 18 January 2006. . Retrieved 10 May 2009.
[142] "ReportDGResearchSocialValuesEN2.PDF" (http://ec.europa.eu/public_opinion/archives/ebs/ebs_225_report_en.pdf) (PDF). . Retrieved 30 October 2010.
[143] "The Duomo of Florence | Tripleman" (http://www.tripleman.com/index.php?showimage=737). www.tripleman.com. . Retrieved 25 March 2010.
[144] "Brunelleschi's Dome" (http://www.brunelleschisdome.com/). Brunelleschi's Dome.com. . Retrieved 25 March 2010.
[145] "The Cardinals of the Holy Roman Church – Living cardinals arranged by country" (http://www.fiu.edu/~mirandas/countrynow.htm#Top). Fiu.edu. 15 October 2010. . Retrieved 30 October 2010.
[146] "Italy – Italian Language, Culture, Customs and Business Etiquette" (http://www.kwintessential.co.uk/resources/global-etiquette/italy-country-profile.html). Kwintessential. . Retrieved 2 August 2010.
[147] "The Holy Orthodox Archdiocese of Italy and Malta" (http://www.ortodossia.it/The Holy Orthodox Archdiocese of Italy ed Malta.htm). Ortodossia – Arcidiocesi Ortodossa d'Italia e Malta. . Retrieved 30 October 2010.
[148] "Le religioni in Italia: I Testimoni di Geova (Religions in Italy: The Jehovah's Witnesses)" (http://www.cesnur.org/religioni_italia/t/testimoni_geova_02.htm) (in Italian). Center for Studies on New Religions. . Retrieved 2011-05-30.
[149] "Chiesa Evangelica Valdese – Unione delle chiese Metodiste e Valdesi (Waldensian Evangelical Church – Union of Waldensian and Methodist churches)" (http://www.chiesavaldese.org/pages/storia/dove_viviamo.php) (in Italian). Chiesa Evangelica Valdese – Unione delle chiese Metodiste e Valdesi (Waldensian Evangelical Church – Union of Waldensian and Methodist churches. . Retrieved 2011-05-30.
[150] "World Council of Churches – Evangelical Methodist Church in Italy" (http://www.oikoumene.org/en/member-churches/regions/europe/italy/evangelical-methodist-church-in-italy.html). World Council of Churches. . Retrieved 30 October 2010.
[151] "Italy: Country's muslims raise funds to help quake victims – Adnkronos Religion" (http://www.adnkronos.com/AKI/English/Religion/?id=3.0.3202304679). Adnkronos. 7 April 2003. . Retrieved 30 October 2010.
[152] "Muslims in Europe: Country guide" (http://news.bbc.co.uk/2/hi/europe/4385768.stm). *BBC News*. 23 December 2005. . Retrieved 5 May 2010.
[153] Rosenthal, Elisabeth (25 July 2005). "Pressure is growingon Muslims in Italy" (http://www.nytimes.com/2005/07/24/world/europe/24iht-rome.html?_r=1). *The New York Times*. . Retrieved 31 March 2010.
[154] "NRI Sikhs in Italy" (http://www.nriinternet.com/EUROPE/ITALY/2004/111604Gurdwara.htm). Nriinternet.com. 15 November 2004. . Retrieved 30 October 2010.
[155] "Unione Buddhista Italiana – UBI: L'Ente" (http://www.buddhismo.it/ente.htm). Buddhismo.it. 18 August 2009. . Retrieved 30 October 2010.
[156] "Most Baha'i Nations (2005)" (http://www.thearda.com/QuickLists/QuickList_40c.asp). *QuickLists > Compare Nations > Religions >*. The Association of Religion Data Archives. 2005. . Retrieved 30 January 2010.
[157] "Il Tempo – È il più grande ateneo d'Europa" (http://iltempo.ilsole24ore.com/2008/09/20/929140-grande_ateneo_europa.shtml). Iltempo.ilsole24ore.com. . Retrieved 30 October 2010.
[158] "Si abbassa l'obbligo scolastico a 15 anni. L'apprendistato diventa scuola" (http://www.flcgil.it/notizie/news/2010/marzo/si_abbassa_l_obbligo_scolastico_a_15_anni_l_apprendistato_diventa_scuola). Flcgil.it. . Retrieved 30 October 2010.
[159] Nancy Imelda Schafer, ISI. "SCI-BYTES: Science in Italy, 1998–2002" (http://in-cites.com/research/2003/june_9_2003-1.html). In-cites.com. . Retrieved 27 January 2010.
[160] "Conferenze, ospiti, news ed eventi legati agli MBA della SDA Bocconi – MBA SDA Bocconi" (http://mba.sdabocconi.it/home/main.php?id=12001&ym=2007-09). SDA Bocconi. . Retrieved 30 October 2010.

[161] "Gatech :: OIE :: GT Study Abroad Programs" (http://web.archive.org/web/20080508020611/http://www.oie.gatech.edu/sa/programs/show.html?id=bocc). Web.archive.org. Archived from the original (http://www.oie.gatech.edu/sa/programs/show.html?id=bocc) on 8 May 2008. . Retrieved 30 October 2010.
[162] "Sda Bocconi supera London Business School – ViviMilano" (http://www.corriere.it/vivimilano/cronache/articoli/2008/05_Maggio/12/sda_bocconi.shtml). Corriere della Serra. 12 November 2008. . Retrieved 30 October 2010.
[163] "Politecnico di milano" (http://www.topuniversities.com/university/408/politecnico-di-milano). Top Universities. . Retrieved 27 October 2009.
[164] "Vision Forum Web Site" (http://www.visionwebsite.eu/vision/progetti_2.php?progetto=28). Vision. . Retrieved 27 October 2009.
[165] "Top 100 European Universities" (http://web.archive.org/web/20080430202044/http://www.arwu.org/rank/2005/ARWU2005_TopEuro.htm). Arwu.org. Archived from the original (http://www.arwu.org/rank/2005/ARWU2005_TopEuro.htm) on 30 April 2008. . Retrieved 27 October 2009.
[166] "ARWU2008" (http://web.archive.org/web/20080822033350/http://www.arwu.org/rank2008/ARWU2008_TopEuro(EN).htm). Arwu.org. Archived from the original (http://www.arwu.org/rank2008/ARWU2008_TopEuro(EN).htm) on 22 August 2008. . Retrieved 27 October 2009.
[167] "Università di Bologna (oldest university in the world)" (http://virtualglobetrotting.com/map/universit-di-bologna-oldest-university-in-the-world/). Virtual Globetrotting. 21 October 2006. . Retrieved 27 October 2009.
[168] "Italy – Health" (http://dev.prenhall.com/divisions/hss/worldreference/IT/health.html). Dev.prenhall.com. . Retrieved 2 August 2010.
[169] "OECD Health Data 2008 How Does Italy Compare" (http://www.oecd.org/dataoecd/45/52/38979929.pdf). OECD. 2008. .
[170] "The World Health Organization's ranking of the world's health systems" (http://www.photius.com/rankings/healthranks.html). ΦΩΤΗΣ ΚΟΥΤΣΟΥΚΗΣ (Photius Coutsoukis). . Retrieved 27 October 2009.
[171] "Health system attainment and performance in all Member States" (http://www.photius.com/rankings/world_health_performance_ranks.html). ΦΩΤΗΣ ΚΟΥΤΣΟΥΚΗΣ (Photius Coutsoukis). . Retrieved 27 October 2009.
[172] United Nations World Population Prospects: 2006 revision (http://www.un.org/esa/population/publications/wpp2006/WPP2006_Highlights_rev.pdf) – Table A.17 for 2005–2010
[173] "Global Prevalence of Adult Obesity" (http://www.webcitation.org/5lwMsu50m) (PDF). *International Obesity Taskforce*. Archived from the original (http://www.iotf.org/database/documents/GlobalPrevalenceofAdultObesity16thDecember08.pdf) on 2009-12-11. . Retrieved 29 January 2008.
[174] World Health Organization (2008) (PDF). *WHO Report on the Global Tobacco Epidemic, 2008: the MPOWER package* (http://www.who.int/entity/tobacco/mpower/mpower_report_full_2008.pdf). World Health Organization. pp. 267–288. ISBN 978-92-4-159628-2. . Retrieved 1 January 2008.
[175] "Smoking Ban Begins in Italy" (http://www.dw-world.de/dw/article/0,,1453590,00.html). Deutsche Welle. 10 January 2005. . Retrieved August 2010.
[176] Eyewitness Travel (2005), pg. 19
[177] "Piazza del Duomo, Pisa – UNESCO World Heritage Centre" (http://whc.unesco.org/en/list/395). Whc.unesco.org. . Retrieved 2011-09-07.
[178] "Leaning Tower of Pisa – Pisa, Italy" (http://www.sacred-destinations.com/italy/pisa-leaning-tower). Pisa, Tuscany, Italy: Sacred-destinations.com. . Retrieved 2011-09-07.
[179] Architecture in Italy (http://www.justitaly.org/italy/italy-architecture.asp), ItalyTravel.com
[180] http://www.bbc.co.uk/history/historic_figures/jones_inigo.shtml
[181] Vitruvian Man is referred to as "iconic" at the following websites and many others: Vitruvian Man (http://www.italian-renaissance-art.com/Vitruvian-Man.html), Fine Art Classics (http://artpassions.com/art/1109-Fine-Art-Classics/0000067329-Leonardo-Da-Vinci-Vitruvian-Man.html), Key Images in the History of Science (http://www.timeshighereducation.co.uk/story.asp?storyCode=403230&sectioncode=26); Curiosity and difference (http://www.ingenious.org.uk/read/identity/bodyimage/Curiosityanddifference/); The Guardian: The Real da Vinci Code (http://www.guardian.co.uk/artanddesign/2006/aug/30/art1)
[182] "All Nobel Prizes in Literature" (http://nobelprize.org/nobel_prizes/literature/laureates/). Nobelprize.org. . Retrieved 2011-05-30.
[183] "Quick Opera Facts 2007" (http://web.archive.org/web/20061001054025/http://www.operaamerica.org/pressroom/quickfacts2006.html). OPERA America. 2007. Archived from the original (http://www.operaamerica.org/pressroom/quickfacts2006.html) on 1 October 2006. . Retrieved 23 April 2007.
[184] Alain P. Dornic (1995). "An Operatic Survey" (http://opera.stanford.edu/misc/Dornic_survey.html). Opera Glass. . Retrieved 23 April 2007.
[185] Kimbell, David R. B (29 April 1994). *Italian Opera – Google Books* (http://books.google.com/?id=C37Gq2GagZIC&dq=Italian+opera&printsec=frontcover&q=). Books.google.co.uk. ISBN 9780521466431. . Retrieved 20 December 2009.
[186] Burke and Waller, 12
[187] "The Cinema Under Mussolini" (http://ccat.sas.upenn.edu/italians/resources/Amiciprize/1996/mussolini.html). Ccat.sas.upenn.edu. . Retrieved 30 October 2010.
[188] "Historical origins of italian neorealism – Neorealism – actor, actress, film, children, voice, show, born, director, son, cinema, scene" (http://www.filmreference.com/encyclopedia/Independent-Film-Road-Movies/Neorealism-HISTORICAL-ORIGINS-OF-ITALIAN-NEOREALISM.html). Filmreference.com. . Retrieved 2011-09-07.

[189] "Italian Neorealism – Explore – The Criterion Collection" (http://www.criterion.com/explore/6-italian-neorealism). Criterion.com. . Retrieved 2011-09-07.
[190] Weidhorn, Manfred (2005). *The Person of the Millennium: The Unique Impact of Galileo on World History*. iUniverse. p. 155. ISBN 0-595-36877-8.
[191] Finocchiaro (2007)
[192] "Galileo and the Birth of Modern Science, by Stephen Hawking, American Heritage's Invention & Technology, Spring 2009, Vol. 24, No. 1, p. 36
[193] "Previous FIFA World Cups™" (http://www.fifa.com/worldcup/archive/index.html). FIFA.com. . Retrieved 8 January 2011.
[194] "The Global Language Monitor " Fashion" (http://www.languagemonitor.com/popular-culture/fashion). Languagemonitor.com. 20 July 2009. . Retrieved 27 October 2009.
[195] http://books.google.com/books?id=pkeaOOxb_isC&pg=PA16&sig=JJmC6YC-Hmmm6Z5P2Cfbvz_jrSA#v=onepage&q=&f=false
[196] Miller (2005) p. 486
[197] Insight Guides (2004) p.220
[198] "Design City Milan" (http://www.wiley.com/WileyCDA/WileyTitle/productCd-0470026839.html). Wiley. . Retrieved 3 January 2010.
[199] "Frieze Magazine – Archive – Milan and Turin" (http://www.frieze.com/issue/article/milan_turin). Frieze. . Retrieved 3 January 2010.
[200] Del Conte, 11–21.

References

Bibliography

- *Italy*. DK. 2005. ISBN 1405307811.
- Miller, Judith (2005). *Furniture: world styles from classical to contemporary*. DK Publishing. ISBN 075661340X.
- *Northern Italy*. APA Publications. 2004. ISBN 9812349030.
- Hacken, Richard. "History of Italy: Primary Documents" (http://eudocs.lib.byu.edu/index.php/History_of_Italy:_Primary_Documents). EuroDocs: Harold B. Lee Library: Brigham Young University. Retrieved 6 March 2010.
- "FastiOnline: A database of archaeological excavations since the year 2000" (http://www.fastionline.org/). International Association of Classical Archaeology (AIAC). 2004-2007. Retrieved 6 March 2010.
- "Italy History – Italian History Index" (http://vlib.iue.it/hist-italy/Index.html) (in Italian, English). European University Institute, The World Wide Web Virtual Library. 1995-2010. Retrieved 6 March 2010.

External links

Country profiles

- by the BBC News (http://news.bbc.co.uk/1/hi/world/europe/country_profiles/1065345.stm)
- by the Economist (http://www.economist.com/countries/Italy/)
- by the U.S. Department of State (http://www.state.gov/p/eur/ci/it/)

General

- Italy (https://www.cia.gov/library/publications/the-world-factbook/geos/it.html) entry at *The World Factbook*
- Italia Tourism official website (http://www.italia.it/en/home.html)
- Italy (http://ucblibraries.colorado.edu/govpubs/for/italy.htm) at *UCB Libraries GovPubs*
- Italy (http://www.dmoz.org/Regional/Europe/Italy/) at the Open Directory Project
- Wikimedia Atlas of Italy
- Geographic data related to Italy (http://www.openstreetmap.org/browse/relation/365331) at OpenStreetMap
- Italy travel guide from Wikitravel
- Key Development Forecasts for Italy (http://www.ifs.du.edu/ifs/frm_CountryProfile.aspx?Country=IT) from International Futures

Government

- President of the Republic of Italy (http://www.quirinale.it/qrnw/statico/lingue/en/en-nap-biografia.htm)
- Parliament (http://www.parlamento.it/) (Italian)
- Chamber of Deputies (http://english.camera.it/)
- Senate (http://www.senato.it/) (Italian)
- Main institutional portal (http://www.italia.gov.it/) (Italian)
- Council of Ministries (http://www.governo.it/) (Italian)
- Constitutional Court (http://www.cortecostituzionale.it/) (Italian)
- Supreme Court (http://www.cortedicassazione.it/) (Italian)
- Ministry of Foreign Affairs (http://www.esteri.gov.it/eng/)
- Ministry of the Interior (http://www.interno.it/mininterno/export/sites/default/en/)
- Ministry of Education (http://www.istruzione.it/) (Italian)
- Ministry of Education: international exchanges (http://www.study-in-italy.it/)
- Ministry of Health (http://www.salute.gov.it/indexEnglish.jsp)
- Ministry of Defense (http://www.difesa.it/) (Italian)
- Ministry of Labour and Social Welfare (http://www.lavoro.gov.it/) (Italian)
- Ministry of Agriculture (http://www.politicheagricole.gov.it/) (Italian)
- Ministry of Justice (http://www.giustizia.gov.it/) (Italian)
- Chief of State and Cabinet Members (https://www.cia.gov/library/publications/world-leaders-1/world-leaders-i/italy.html)

Public institutions

- ENIT North America (http://www.italiantourism.com/)
- Italian railways (http://www.trenitalia.com/cms/v/index.jsp?vgnextoid=ad1ce14114bc9110VgnVCM10000080a3e90aRCRD)
- Italian national and regional parks (http://www.parks.it/Eindex.html)

Ticino_(river)

Ticino	
 The river Ticino and the Old Bridge of Pavia (medieval in date)	
Origin	Val Bedretto, Ticino, Switzerland
Mouth	River Po, south-east of Pavia, Italy
Basin countries	Switzerland, Italy
Length	248 km (**unknown operator: u'strong'** mi)
Source elevation	about 2200 m (**unknown operator: u'strong'** ft)
Avg. discharge	at Magadino: 69 m^3/s
Basin area	7228 km^2 (**unknown operator: u'strong'** sq mi)

The river **Ticino** (Lombard: *Tisín*; German: *Tessin*; Latin: *Ticinus*) is a left-bank tributary of the Po River. It has given its name to the Swiss canton through which its upper portion flows.

The course

The river rises in the Val Bedretto in Switzerland at the frontier between the cantons of Valais and Ticino and later flows through Lake Maggiore, before entering Italy. The Ticino joins the Po a few kilometres downstream (along the Ticino) from Pavia. It is about 248 kilometres (**unknown operator: u'strong'** mi) long. The highest point of the drainage basin is the summit of Grenzgipfel (a subpeak of Monte Rosa), at 4618 metres (**unknown operator: u'strong'** ft). Beneath it flows the Anza River, a right-bank tributary of the Ticino.

The river is dammed in Switzerland in order to create hydroelectricity, while in Italy it is primarily used for irrigation.

History

Ticino was the location of the Battle of Ticinus, the first battle of the Second Punic War fought between the Carthaginian forces of Hannibal and the Romans under Publius Cornelius Scipio in November 218 BC, and the first battle on Italian soil.

Towns and tributaries along the Ticino

Val Bedretto

Val Bedretto, a narrow alpine valley named after the village of Bedretto, culminates in Nufenen Pass, Italian Passo della Novena, at 2478 m (**unknown operator: u'strong'** ft) located between Pizzo Gallina and Nufenstock. Through it runs the border between the cantons of Valais and Ticino. A road constructed in 1964 goes over the pass ultimately leading to the Rhone river valley on the other side. French is mainly spoken in Valais but near the pass German is spoken. The population of Ticino speaks Italian.

The upper Ticino near Airolo

The road up the val is fairly straight until it approaches the pass, where it becomes hairpin. On the slope below the first hairpin at about 2200 m (**unknown operator: u'strong'** ft) is an area of springs from which the Ticino river originates. Subsequently it becomes a mountain brook flowing straight down the val to the village of All'Acqua or All'Acqua Ospizio at 1614 m (**unknown operator: u'strong'** ft), named for the hospice for travellers located there at the previous end of the road in former centuries. Currently it is a base for skiing and hiking. Below it is Ronco at 1487 m (**unknown operator: u'strong'** ft) and Bedretto at 1402 m (**unknown operator: u'strong'** ft).[1] The val is subject to avalanches and snow can remain on the ground as late as September.

Below Bedretto the Ri di Cristallina, "Cristallina stream", comes in from the right at Ossasco, and further down Fontana, still in Bedretto. The entire area is laced with hiking trails and mountain huts.

Valle Leventina

- Airolo

Other

- in Switzerland: Bellinzona, Locarno (on Lake Maggiore).
- in Italy: Stresa (on Lake Maggiore), Vigevano, Pavia.

References

[1] Reynolds, Kev (1992). "Val Bedretto". *Walking in Ticino, Switzerland: A Walking Guide* (illustrated ed.). Milnthorpe: Cicerone Press Limited. pp. 32–43. ISBN 1852840986, 9781852840983.

Pavia

Pavia	
— Comune —	
Città di Pavia	

 **Coat of arms**	
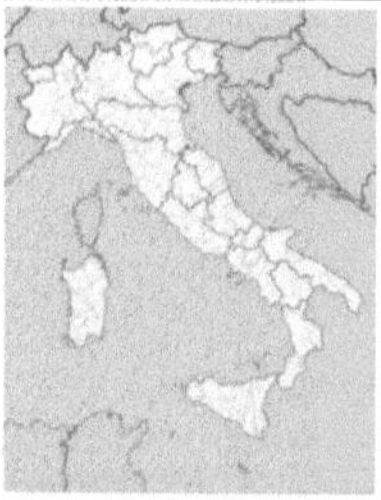 Pavia — Location of Pavia in Italy	
Coordinates: 45°11′N 09°09′E	
Country	Italy
Region	Lombardy
Province	Pavia (PV)
Frazioni	Albertario, Ca' della Terra, Cantone Tre Miglia, Cassinino, Cittadella, Fossarmato, Mirabello, Montebellino, Pantaleona, Prado, Villalunga
Government	

• Mayor	Alessandro Cattaneo (PdL)
Area	
• Total	62 km^2 (**unknown operator: u'strong'** sq mi)
Elevation	77 m (**unknown operator: u'strong'** ft)
Population (31 October 2010)	
• Total	71227
• Density	**unknown operator: u'strong'**/km^2 (**unknown operator: u'strong'**/sq mi)
Demonym	Pavesi
Time zone	CET (UTC+1)
• Summer (DST)	CEST (UTC+2)
Postal code	27100
Dialing code	0382
Patron saint	St. Sirus
Saint day	December 9
Website	Official website [1]

Pavia (Italian: [pa'viːa] (listen); Lombard *Pavia*), the ancient Ticinum, is a town and *comune* of south-western Lombardy, northern Italy, 35 km south of Milan on the lower Ticino river near its confluence with the Po. It is the capital of the province of Pavia. It has a population of c. 71,000. The city was also the capital of the Kingdom of the Lombards from 568 to 774.

The Certosa of Pavia.

Pavia is the capital of a fertile eponymous province known for agricultural products including wine, rice, cereals, and dairy products. Although there are a number of industries located in the suburbs, these tend not to disturb the peaceful atmosphere of the town. The town also is home to the ancient University of Pavia. The University, together with the IUSS (Institute for Advanced Studies of Pavia), the Ghislieri College, the Borromeo College, the Nuovo College, the Santa Caterina College and the EDiSU, belongs to the Pavia Study System. Furthermore, Pavia is the see city of the Roman Catholic diocese of Pavia. The city possesses a vast amount of artistic and cultural treasures, including several important churches and museums, such as the well-known Certosa di Pavia.

History

San Michele Maggiore.

Dating back to pre-Roman times, the town of Pavia (then known as *Ticinum*) was a municipality and an important military site (a castrum) under the Roman Empire. It was said by Pliny the Elder to have been founded by the Laevi and Marici, two Ligurian tribes, while Ptolemy attributes it to the Insubres. The Roman city most likely began as a small military camp, built by the consul Publius Cornelius Scipio in 218 BC to guard a wooden bridge he had built over the river Ticinum, on his way to search for Hannibal, who was rumoured to have managed to lead an army over the Alps and into Italy. The forces of Rome and Carthage ran into each other soon thereafter, and the Romans suffered the first of many crushing defeats at the hands of Hannibal, with the consul himself almost losing his life. The bridge was destroyed, but the fortified camp, which at the time was the most forward Roman military outpost in the Po Valley, somehow survived the long Second Punic War, and gradually evolved into a garrison town.

Side view of the Visconti Castle.

Its importance grew with the extension of the Via Aemilia from Ariminum (Rimini) to the Po River (187 BC), which it crossed at Placentia (Piacenza) and there forked, one branch going to Mediolanum (Milan) and the other to Ticinum, and thence to Laumellum where it divided once more, one branch going to Vercellae - and thence to Eporedia and Augusta Praetoria - and the other to Valentia - and thence to Augusta Taurinorum (Turin) or to Pollentia.

Here, in 476, Odoacer defeated Flavius Orestes after a long siege. To punish the city for helping the rival, Odoacer destroyed it completely. However, Orestes was able to escape to Piacenza, where Odoacer followed and killed him, deposing his son Romulus Augustus. This is commonly considered the end of the Western Roman Empire.

A late name of the city in Latin was *Papia* (probably related to the Pope), which evolved to the Italian name *Pavia*. Sometimes it's been referred to as *Ticinum Papia*, combining both Latin names.

Under the Ostrogoths, Pavia became a fortified citadel and their last bulwark in the war against Belisarius.

After the Lombards conquest, Pavia became the capital of their kingdom (568-774). During the Rule of the Dukes, it was ruled by Zaban. It continued to function as the administrative centre of the kingdom, but by the reign of Desiderius, it had deteriorated as a first-rate defensive work and Charlemagne took it in the Siege of Pavia (June, 774) assuming the kingship of the Lombards. The Hungarians burned Pavia from sometime during 889 to 955 Ad. Pavia remained the capital of the Italian Kingdom and the centre of royal coronations until the diminution of imperial authority there in the 12th century. In 1004 Henry II, Holy Roman Emperor bloodily suppressed a revolt of the citizens of Pavia, who disputed his recent crowning as King of Italy.

Santa Maria del Carmine.

In the 12th century Pavia acquired the status of a self-governing commune. In the political division between Guelphs and Ghibellines that characterizes the Italian Middle Ages, Pavia was traditionally Ghibelline, a position that was as much supported by the rivalry with Milan as it was a mark of the defiance of the Emperor that led the Lombard League against the emperor Frederick Barbarossa, who was attempting to reassert long-dormant Imperial influence over Italy.

Cathedral of Pavia.

In the following centuries Pavia was an important and active town. Under the Treaty of Pavia, Emperor Louis IV granted during his stay in Italy the Palatinate to his brother Duke Rudolph's descendants. Pavia held out against the domination of Milan, finally yielding to the Visconti family, rulers of that city in 1359; under the Visconti Pavia became an intellectual and artistic centre, being the seat from 1361 of the University of Pavia founded around the nucleus of the old school of law, which attracted students from many countries.

The Battle of Pavia (1525) marks a watershed in the city's fortunes, since by that time, the former cleavage between the supporters of the Pope and those of the Holy Roman Emperor had shifted to one between a French party (allied with the Pope) and a party supporting the Emperor and King of Spain Charles V. Thus during the Valois-Habsburg Italian Wars, Pavia was naturally on the Imperial (and Spanish) side. The defeat and capture of king Francis I of France during the battle ushered in a period of Spanish occupation which lasted until 1713 at the conclusion of the War of the Spanish Succession. Pavia was then ruled by the Austrians until 1796, when it was occupied by the French army under Napoleon. During this Austrian period the University was greatly supported by Maria Theresa of Austria and saw a great renaissance that eventually led to a second renaissance due to the presence of leading scientists and humanists like Ugo Foscolo, Alessandro Volta, Lazzaro Spallanzani, Camillo Golgi among others.

In 1815, it again passed under Austrian administration until the Second War of Italian Independence (1859) and the unification of Italy one year later.

Main sights

Pavia's most famous landmark is the *Certosa*, or Carthusian monastery, founded in 1396 and located eight kilometres north of the city.

Among other notable structures are:

- Cathedral of Pavia (*Duomo di Pavia*), begun in 1488; however, only by 1898 were the façade and the dome completed according to the original design. The central dome has an octagonal plan, stands 97 m high, and weighs some 20,000 tons. This dome is the third for size in Italy, after St. Peter's Basilica and Santa Maria del Fiore in Florence. Next to the Duomo were the Civic Tower (existing at least from 1330 and enlarged in 1583 by Pellegrino Tibaldi): its fall on March 17, 1989, was the final motivating force that started the last decade's efforts to save the Leaning Tower of Pisa from a similar fate.
- *San Michele Maggiore* (St. Michael) is an outstanding example of Lombard-Romanesque church architecture in Lombardy. It is located on the site of a pre-existing Lombard church, which the lower part of the campanile belongs to. Destroyed in 1004, the church was rebuilt from around the end of the 11th century (including the crypt, the transept and the choir), and finished in 1155. It is characterized by an extensive use of sandstone and by a very long transept, provided with a façade and an apse of its own. In the church the Emperor Frederick Barbarossa was crowned in 1155.
- The Basilica of *San Pietro in Ciel d'Oro* ("St. Peter in Golden Sky"), where Saint Augustine, Boethius and the Lombard king Liutprand are buried, was begun in the 6th century. The current construction was built in 1132. It is similar to San Michele Maggiore, but different in the asymmetric façade with a single portal, the use of brickwork instead of sandstone, and, in the interior, the absence of matronei, galleries reserved for women and the shortest transept. The noteworthy arch housing the relics of St. Augustine was built in 1362 by artists from Campione, and is decorated by some 150 statues and reliefs. The church is mentioned by Dante Alighieri in the X canto of his *Divine Comedy*.
- *San Francesco d'Assisi*, a late Romanesque church (1238–98) with a restored Gothic façade, located on Corso Cairoli.
- *San Teodoro* (1117), dedicated to Theodore of Pavia, a medieval bishop of the Diocese of Pavia, is the third romanesque basilica in the city, though smaller than the former ones. It is situated on the slopes leading down to the Ticino river and served the fishermen. The apses and the three-level tiburium are samples of the effective simplicity of romanesque decoration. Inside are two outstanding bird's-eye-view frescoes of the city (1525) attributed to the painter Bernardino Lanzani. The latter, the definitive release, was stripped off disclosing the unfinished first one. Both are impressively detailed and reveal how little Pavia's urban design has changed during the last 500 years.
- The large fortified *Castello Visconteo* (built 1360-1365 by Galeazzo II Visconti). In spite of its being fortified, it actually was used as a private residence rather than a stronghold. The poet Francesco Petrarca spent some time there, when Gian Galeazzo Visconti called him to take charge of the magnificent library which owned about a thousand books and manuscripts, subsequently lost. The Castle is now home to the City Museums (Musei Civici) and the park is a popular attraction for children. An unconfirmed legend wants the Castle to be connected by a secret underground tunnel to the *Certosa*.
- The church of *Santa Maria del Carmine* is one of the best known examples of Gothic brickwork architecture in northern Italy. It is the second largest church in the city after the cathedral and is built on the Latin cross plan, with a perimeter of 80 x 40 meters comprising a nave and two aisles. The characteristic façade has a large rose window and seven cusps.
- The renaissance church of *Santa Maria di Canepanova* is attributed to Bramante.
- The medieval towers still shape the town skyline. The main clusters still rising are rallied in Piazza Leonardo da Vinci, Via Luigi Porta, and Piazza Collegio Borromeo.

Universities, colleges and other institutions

Pavia is a major Italian college town, with several institutes, universities and academies, including the ancient University of Pavia. Here is an incomplete list of the main institutions located in the city:

- The University of Pavia, one of the most ancient universities in Europe, was founded in 1361, although a school of rhetoric is documented in 825 making this center perhaps the oldest protouniversity of Europe. The *Centrale* Building is a wide block made up of twelve courts of the 15th-19th centuries. The sober façade shifts from baroque style to neoclassic. The *Big Staircase*, the *Aula Foscolo*, the *Aula Volta*, the *Aula Scarpa* and the *Aula Magna* are neoclassic too. The *Cortile degli Spiriti Magni* hosts the statues of some of the most important scholars and alumni. Ancient burial monuments and gravestones of scholars of the 14th-16th centuries are walled up in the *Cortile Voltiano* (most come from demolished churches). The *Cortile delle Magnolie* holds an ancient pit. The *Cortile di Ludovico il Moro* has a renaissance loggia and terracotta decorations. Both courts, as well as two more, were the cloisters of the ancient Ospedale di San Matteo. The Orto Botanico dell'Università di Pavia is the university's botanical garden.
- The Borromeo College (Ital. *Almo Collegio Borromeo*), founded in 1561 by Carlo Borromeo, is the oldest college at the University of Pavia in northern Italy.
- The Ghislieri College (Ital. *Collegio Ghislieri*), founded in 1567 by Pope Pius V, is the second ancient college in Pavia, with the other first being Almo Collegio Borromeo, and one of the most ancient colleges in Italy and co-founder of the IUSS, located in Pavia as well. Collegio Ghislieri is a 450 years old Italian institution committed to promote University studies on the basis of merit, hosting around 200 pupils (males and females) who attend all faculties in Pavia State University, offering them logistic and cultural opportunities such as scholarships, lectures, conferences, a 100,000 volumes library (the third one among private libraries in Northern Italy), foreign languages courses. Each year about 30 new students coming from all over the country are selected by public contest. Founded by Pope Pius V (Antonio Ghislieri) in 1567, since 18th century laically managed, nowadays under the High Patronage of the Presidency of the Italian Republic, it is ranked among high qualifying institutions by the Italian Ministry for Education and University.
- The IUSS Pavia or the "Istituto Universitario di Studi Superiori" of Pavia (Eng. *IUSS School for Advanced Studies*) is a higher learning institute located in Pavia, Italy. It was founded in 1997 by the University of Pavia, the Borromeo College and the Ghislieri College, supported by the Italian Minister of Education. It is shaped according to the Scuola Normale Superiore di Pisa model and reunites all the five colleges of Pavia, forming the Pavia Study System.

Transport

Pavia railway station, opened in 1862, forms part of the Milan–Pavia–Genoa railway, and is also a terminus of four secondary railways, linking Pavia with Alessandria, Cremona, Vercelli and Piacenza.

People

People born in Pavia include:

- Lanfranc (c. 1005–1089), abbot and Archbishop of Canterbury
- Gerolamo Cardano (1501–1576), scientist
- Benedetto Cairoli (1825–1889), twice head of the government
- Francesco Corbetta (1615–1681), guitar virtuoso, teacher and composer
- Tranquillo Cremona (1837–1878), painter
- Claudia Muzio (1889–1936), opera singer
- Carlo M. Cipolla (1922–2000), economic historian

The Collegio Borromeo.

People who have lived in Pavia include:

- St. Alexander Sauli (1591–1592), Bishop of Pavia
- Alessandro Volta (1745–1827), scientist
- Simion Bărnuţiu (1808–1864), philosopher and politician
- Camillo Golgi (1843–1926), biologist and Nobel laureate
- Giovanni de Ventura (fl. 1479), fifteenth century plague doctor
- Riccardo Pampuri (1897–1930), saint and medical doctor
- Albert Einstein (1879–1955), physicist and Nobel laureate
- Ugo Foscolo (1778–1827), Italian writer, revolutionary and poet.
- Dante Troisi (1920–1989), writer and judge[2]

The University of Pavia's Aula Magna.

Among the illustrious scholars who studied or taught at the University of Pavia, the following are at least worth remembering: Carlo Goldoni, Gerolamo Cardano, Gerolamo Saccheri, Ugo Foscolo, Alessandro Volta the inventor of the battery, Lazzaro Spallanzani, Antonio Scarpa, Carlo Forlanini and the Nobel laureate biologist Camillo Golgi.

References

[1] http://http://www.comune.pavia.it

[2] Pallotta, Augustus (1999). *Italian Novelists Since World War II, 1965-1995* (http://books.google.com/books?id=wdEUAQAAIAAJ&dq=Dante+Troisi). Dictionary of Literary Biography. **196**. Detroit: Gale Research. .

External links

- Pavia Touristic Promotion (http://www.provincia.pv.it/provinciapv/brick/turismo)
- Pavia Touristic Portal (http://www.paviafree.it/)
- Oltrepò pavese (http://www.oltrepeat.com/it/oltrepo-pavese.html)
- Pavia pictures (http://www.fotoadamo.net/adamo/categorie/index.asp?vs=3&c=96&s=3&ca=1)
- Images of Pavia (http://www.flickr.com/groups/pavia_in_immagini)
- Weather in Pavia (http://www.paviameteo.it/)

Italian_language

Italian	
Italiano, Lingua italiana	
Pronunciation	Italian pronunciation: [itaˈljaːno]
Spoken in	• European Union • Italy • San Marino • Sovereign Military Order of Malta • Switzerland • Vatican City • Slovenia (only in Slovenian Istria) • Croatia (only in Istria County) • Somalia (Transitional Federal Parliament) • Eritrea Although Eritrea has no official language, Italian is still well-diffused among older people and in administrative, commercial and teaching-related areas. • Libya Although not official, Italian is still widely known among older populations and is used in the commercial and education sectors. • Malta • Kosovo • Montenegro
Native speakers	62 million Italian proper, native and native bilingual (no date) Total: 80 million;[1] 85 million all varieties[1]
Language family	Indo-European • Italic • Romance • Italo-Dalmatian • **Italian**
Writing system	Latin (Italian alphabet)
Official status	
Official language in	European Union Italy Switzerland San Marino Vatican City Croatia (Istria County) Slovenia (Slovenian Istria)
Regulated by	not officially by Accademia della Crusca
Language codes	
ISO 639-1	`it`
ISO 639-2	`ita`
ISO 639-3	`ita`
Linguasphere	`51-AAA-q`

Italian (*italiano* or *lingua italiana*) is a Romance language spoken mainly in Europe: Italy, Switzerland, San Marino, Vatican City, by minorities in Malta, Monaco, Croatia, Slovenia, France, Libya, Eritrea, and Somalia,[2] and by immigrant communities in the Americas and Australia. Many speakers are native bilinguals of both standardised Italian and other regional languages.[3]

According to the Bologna statistics of the European Union, Italian is spoken as a mother tongue by 65 million people in the EU (13% of the EU population), mainly in Italy, and as a second language by 14 million (3%).[1] Including the Italian speakers in non-EU European countries (such as Switzerland and Albania) and on other continents, the total number of speakers is more than 85 million.

In Switzerland, Italian is one of four official languages; it is studied and learned in all the confederation schools and spoken, as mother tongue, in the Swiss cantons of Ticino and Grigioni and by the Italian immigrants that are present in large numbers in German- and French-speaking cantons. It is also the official language of San Marino, as well as the primary language of Vatican City.[4] It is co-official in Slovenian Istria and in part of the Istria County in Croatia. The Italian language adopted by the state after the unification of Italy is based on the Tuscan dialect, which beforehand was a language spoken mostly by the upper class of Florentine society.[5] Its development was also influenced by other Italian dialects and by the Germanic language of the post-Roman invaders.

Italian derives diachronically from Latin. Unlike most other Romance languages, Italian retains Latin's contrast between short and long consonants. As in most Romance languages, stress is distinctive. In particular, among the Romance languages, Italian is the closest to Latin in terms of vocabulary.[6]

History

The standard Italian language has a poetic and literary origin starting in the twelfth century, and the modern standard of the language was largely shaped by relatively recent events. However, Italian as a language used in the Italian Peninsula has a longer history. In fact the earliest surviving texts that can definitely be called Italian (or more accurately, vernacular, as distinct from its predecessor Vulgar Latin) are legal formulae from the Province of Benevento that date from 960–963.[7] What would come to be thought of as Italian was first formalized in the early fourteenth century through the works of Tuscan writer Dante Alighieri, written in his native Florentine. Dante's epic poems, known collectively as the *Commedia,* to which another Tuscan poet Giovanni Boccaccio later affixed the title *Divina,* were read throughout Italy and his written dialect became the "canonical standard" that all educated Italians could understand. Dante is still credited with standardizing the Italian language, and thus the dialect of Florence became the basis for what would become the official language of Italy.

Italian often was an official language of the various Italian states predating unification, slowly usurping Latin, even when ruled by foreign powers (such as the Spanish in the Kingdom of Naples, or the Austrians in the Kingdom of Lombardy-Venetia), even though the masses spoke primarily vernacular languages and dialects. Italian was also one of the many recognised languages in the Austro-Hungarian Empire.

Italy has always had a distinctive dialect for each city, since the cities, until recently, were thought of as city-states. Those dialects now have considerable variety. As Tuscan-derived Italian came to be used throughout Italy, features of local speech were naturally adopted, producing various versions of Regional Italian. The most characteristic differences, for instance, between Roman Italian and Milanese Italian are the gemination of initial consonants and the pronunciation of stressed "e", and of "s" in some cases: e.g. *va bene* "all right": is pronounced [va ˈbːɛne] by a Roman (and by any standard-speaker, like a Florentine), [va ˈbene] by a Milanese (and by any speaker whose native dialect lies to the north of La Spezia-Rimini Line); *a casa* "at home" is [a ˈkːasa] for Roman and standard, [a ˈkaza] for Milanese and generally northern.

In contrast to the Northern Italian language, southern Italian dialects and languages were largely untouched by the Franco-Occitan influences introduced to Italy, mainly by bards from France, during the Middle Ages but, after the Norman conquest of southern Italy, Sicily became the first Italian land to adopt Occitan lyric moods (and words) in poetry. Even in the case of Northern Italian language, however, scholars are careful not to overstate the effects of outsiders on the natural indigenous developments of the languages.

The economic might and relatively advanced development of Tuscany at the time (Late Middle Ages) gave its dialect weight, though the Venetian language remained widespread in medieval Italian commercial life, and Ligurian (or Genoese) remained in use in maritime trade alongside the Mediterranean. The increasing political and cultural relevance of Florence during the periods of the rise of Medici's bank, Humanism, and the Renaissance made its dialect, or rather a refined version of it, a standard in the arts.

Renaissance

Starting with the Renaissance Italian became the language used in the courts of every state in the peninsula. The rediscovery of Dante's *De vulgari eloquentia* and a renewed interest in linguistics in the sixteenth century, sparked a debate that raged throughout Italy concerning the criteria that should govern the establishment of a modern Italian literary and spoken language. Scholars divided into three factions:

- The purists, headed by Venetian Pietro Bembo (who, in his *Gli Asolani*, claimed the language might be based only on the great literary classics, such as Petrarch and some part of Boccaccio). The purists thought the Divine Comedy not dignified enough, because it used elements from non-lyric registers of the language.
- Niccolò Machiavelli and other Florentines preferred the version spoken by ordinary people in their own times.
- The courtiers, like Baldassare Castiglione and Gian Giorgio Trissino, insisted that each local vernacular contribute to the new standard.

A fourth faction claimed the best Italian was the one that the papal court adopted, which was a mix of Florentine and the dialect of Rome. Eventually, Bembo's ideas prevailed, and the foundation of the Accademia della Crusca in Florence (1582–1583), the official legislative body of the Italian language led to publication of Agnolo Monosini's Latin tome Floris italicae linguae libri novem in 1604 followed by the first Italian dictionary in 1612.

Modern era

An important event that helped the diffusion of Italian was the conquest and occupation of Italy by Napoleon in the early nineteenth century (who was himself of Italian-Corsican descent). This conquest propelled the unification of Italy some decades after, and pushed the Italian language into a lingua franca used not only among clerks, nobility and functionaries in the Italian courts but also in the bourgeoisie.

Contemporary times

Italian literature's first modern novel, *I Promessi Sposi* (*The Betrothed*), by Alessandro Manzoni, further defined the standard by "rinsing" his Milanese "in the waters of the Arno" (Florence's river), as he states in the Preface to his 1840 edition.

After unification a huge number of civil servants and soldiers recruited from all over the country introduced many more words and idioms from their home languages ("ciao" is derived from Venetian word "s-cia[v]o" (*slave*), "panettone" comes from Lombard word "panatton" etc.). Only 2.5% of Italy's population could speak the Italian standardized language properly when the nation unified in 1861.[8]

Classification

Italian is a Romance language; it derives diachronically from Latin. It is part of the Italic branch of the Indo-European language family. Italian is related most closely to the other two Italo-Dalmatian languages, Sicilian and the extinct Dalmatian.

Unlike most other Romance languages, Italian retains Latin's contrast between short and long consonants. As in most Romance languages, stress is distinctive. In particular, among the Romance languages, Italian is the closest to Latin in terms of vocabulary.[6] Lexical similarity is 89% with French, 87% with Catalan, 85% with Sardinian, 82% with Spanish and Portuguese, 78% with Rhaeto-Romance, and 77% with Romanian.[2] [9]

Geographic distribution

The list below shows the geographical distribution of the Italian language around the world. The total number of native speakers of Italian is 62 million people according to Encarta[10] and Ethnologue.[2] But in the statistics of the European Union, Italian is spoken as a mother tongue by 13% of the EU population or 65 million people, mainly in Italy. Also in the EU, it is spoken as a second language by 3% of the population or by 14 million people.[1] Including the Italian speakers in non-EU European countries (such as Switzerland and Albania) and on other continents (especially in Argentina, Brazil, the US, Canada, Australia, Venezuela, as is shown below), the total number of speakers is more than 85 million.

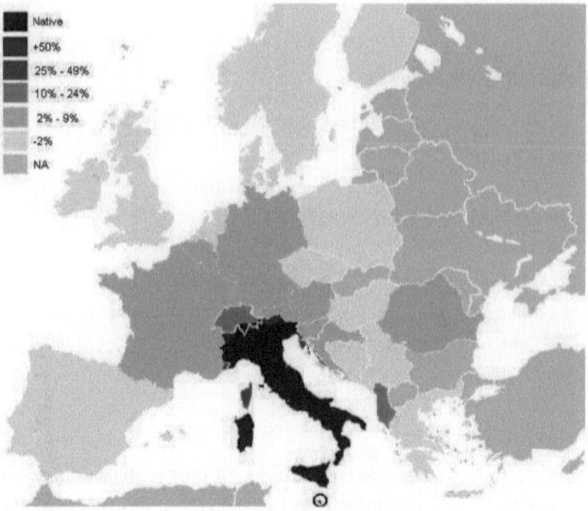

Knowledge of Italian in Europe

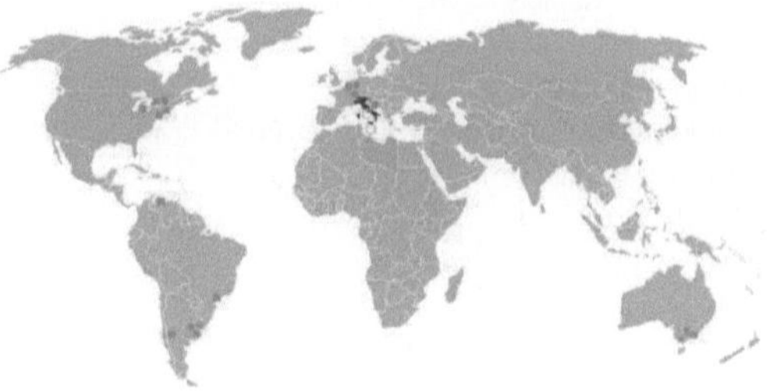

The geographic distribution of the Italian language in the world: large Italian-speaking communities are shown in green; light blue indicates areas where the Italian language was used officially during the Italian colonial period.

Official

- European Union
- Italy
- San Marino
- Sovereign Military Order of Malta
- Switzerland
- Vatican City
- Slovenia (only in Slovenian Istria)
- Croatia (only in Istria County)

Secondary

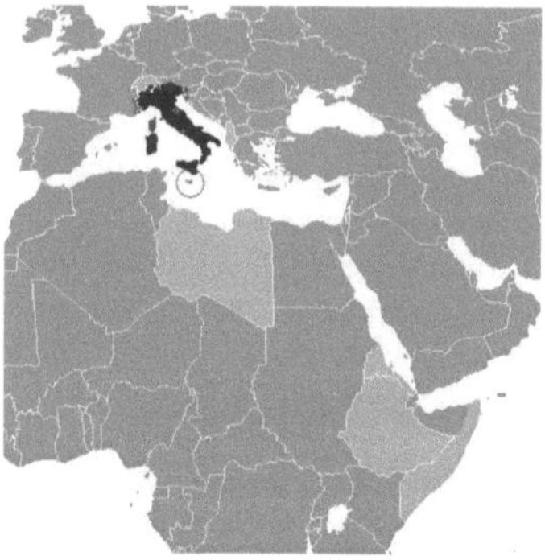

Use of the Italian language in Europe and Africa

- Somalia (Transitional Federal Parliament)
- Eritrea Although Eritrea has no official language, Italian is still well-diffused among older people and in administrative, commercial and teaching-related areas.
- Libya Although not official, Italian is still widely known among older populations and is used in the commercial and education sectors.
- Malta
- Kosovo
- Montenegro

Historically significant

- France (in Corsica, Savoy, Nice and some valleys)
- Albania
- Croatia (Istria, Rijeka, Kvarner, Dalmatia)
- Slovenia (Slovenian Littoral)
- Malta
- Monaco
- Montenegro
- Greece (in Dodecanese 1912–1943)
- Ukraine (in Crimea[11])

Historically official

- Kingdom of Italy (1861–1946 and until 1861 in all the former Italian states before the unification; also in Italian Social Republic 1943–1945 and in Free Territory of Trieste 1947–1954)
- Croatia (in Istria X century-1797, Dalmatia XII century-1797, Rijeka, Zadar, Lastovo and Palagruža 1919–1947, and in the Governorship of Dalmatia 1941–1943)
- Trieste (Free Territory of Trieste 1947-1954)
- Fiume (Free State of Fiume 1920–1924)
- Eritrea (1890–1941)
- Somalia (Italian Somaliland 1895–1960, British Somaliland 1940–1941)
- Ethiopia (Abissinia 1936–1941)
- Cyprus (1489–1571)
- China (in Tientsin 1901–1944)
- Libya (1911–1943)
- Egypt (in Western part 1940–1942)
- Montenegro (Kingdom of Montenegro (1941–1944))
- Slovenia (in the Slovenian Littoral, western Inner Carniola 1919–1947, and in the Province of Ljubljana 1941–1943)
- Greece (in Crete XIII century-1699; in the Venetian Ionian Islands 1204-1797 and 1941–1943; in Chios 1261-1566; in the Dodecanese 1912–1943 and in many other islands and cities under the Venetian and Genoese domination between the XII and XVI centuries)
- Albania (in Venetian Albania 1420-1797; in Sazan Island 1920–1947 and in all the country 1938–1945)
- Malta (until 1934)

- France (in Tende, La Brigue and other small valleys until 1947, in Corsica until 1895 and between 1942–1943 and in the territory of Nice X century-1860 and 1942–1943)
- Tunisia (in Tabarka 1540-1742 and in all the country 1942–1943)
- Ukraine (in Crimea under the Genoese domination between the XIII and XV centuries)
- Austria-Hungary (in the Austrian Littoral, Fiume, Dalmatia and Trentino until 1918)
- Turkey (in the district of Pera 1204-1453 and in Territory of Antalya 1919–1922)

Used by some immigrant communities

- Argentina 1,500,000[12]
- United States 1,008,370[13]
- France 500,000–1,000,000
- Canada 661,000[14]
- Germany 548,000[15]
- Switzerland over 500,000 (mentioned here because related to the German and French speaking areas)
- Venezuela 400,000[16]
- Australia 353,605[17]
- Belgium 250,000
- Mexico 220,000
- United Kingdom 200,000[18]
- Uruguay 79,000[19]
- Egypt 72,400[20]
- Brazil 50,000[21] (the dominant immigrant language in Brazil, Talian, is of Venetian origin, and spoken by c. 4,000,000 people)[21]
- Colombia 2,250[22]

Italian is the official language of Italy and San Marino, and one of the official languages of Switzerland, spoken in the cantons of Ticino and part of Graubünden (Grigioni in Italian), which together are a region referred to as Italian Switzerland. It is also the official language with Croatian and Slovenian in some areas of Istria, where an Italian minority exists. In the Brazilian cities of Santa Teresa and Vila Velha it enjoys official status alongside Portuguese, being "knighted" as an ethnic language. It is the primary language of the Vatican City and is widely used and taught in Monaco and Malta. It served as Malta's official language until the Maltese language was enshrined in the 1934 Constitution. It is also spoken to a significant extent in France, with over 1,000,000 speakers [23] (especially in Corsica and the County of Nice, areas that historically spoke Italian dialects before annexation to France), and it is understood by large parts of the populations of Albania, coastal Montenegro and western Slovenia, reached by many Italian television channels.

Italian is also spoken by some in former Italian colonies in Africa (Libya and Eritrea). However, its use has dropped sharply since the colonial period. In Eritrea, Italian is widely understood.[24] In fact, for fifty years, during the colonial period, Italian was the language of education, but as of 1997, there is only one Italian-language school remaining, with four hundred seventy pupils yearly. The name of the only Italian-language school in Eritrea is Scuola Italiana di Asmara,[25] which was also the only Italian-language school in Ethiopia, when Eritrea was a province of Ethiopia.[26] The number of Italian speakers may increase a little when the number of students at that school increases and because it is still spoken in commerce,[27] and Eritrea will be the only African nation where Italian is widely spoken and understood. In Libya, since 1969, Italian has been wiped out by the Libyan Revolution's Arabization programmes in education and media. In Egypt and Tunisia, it is spoken mostly by Italian Egyptians, Italian Tunisians, and some professionals of non-Italian descent. In all of the above former Italian African colonies, most of the fluent Italian speakers are people who grew up in officially Italian-speaking nations, especially Italy, and returned to Africa.

Italian and Italian dialects are widely used by Italian immigrants and many of their descendants living throughout Western Europe (especially France, Germany, Belgium, Switzerland, the United Kingdom and Luxembourg), the United States, Canada, Australia, and Latin America (especially Uruguay, Brazil, Argentina, and Venezuela).

In the United States, the largest Italian-speaking populations are found in five cities: Boston (7,000),[28] Chicago (12,000),[29] the Miami region (27,000),[30] New York City (140,000),[31] and Philadelphia (15,000).[32] According to the United States Census in 2000, over 1 million Italian Americans spoke Italian at home, with the largest concentrations—and nearly half of the total—found in the states of New York (294,271) and New Jersey (116,365).[33] In Canada, Italian is the fourth most commonly spoken language, with 661,000 speakers (or about 2.1% of the population) according to the 2006 Census. Particularly large Italian-speaking communities are found in Montreal (c. 179,000) and Toronto (c. 262,000).[14] Italian is also strongly visible in the Hamilton area. Italian is the second most commonly spoken language in Australia, where 353,605 Italian Australians, or 1.9% of the population, reported speaking Italian at home in the 2001 Census.[34] In 2001 there were 130,000 Italian speakers in Melbourne,[35] and 90,000 in Sydney.[36]

Education

Italian is widely taught in many schools around the world, but rarely as the first foreign language; in fact, Italian is considered the fourth- or fifth - most taught foreign language in the world.[37]

According to the Italian Ministry of Foreign Affairs, every year there are more than 200,000 foreign students that are learning Italian language; they are distributed in the 90 Institutes of Italian Culture in the world, in the 179 Italian schools abroad and in the 111 Italian sections that are open into foreign schools.[38]

In the United States, Italian is the fourth most taught foreign language after Spanish, French and German, in that order (the fifth, considering also the American Sign Language).[39] In the anglophone Canada Italian is second after French but in the United Kingdom it is the fourth after French, Spanish and German.[40] In central-east Europe Italian is first in Albania and Montenegro, second in Austria, Croatia, Slovenia, and Ukraine after English, and third in Hungary, Romania and Russia after English and German.[38] But throughout the world, Italian is the fifth most taught foreign language, after English, Spanish, French, and German.[41]

In the European Union statistics, Italian is spoken as a mother tongue by 13% of the population or 65 million people,[1] mainly in Italy. In the EU, it is spoken as a second language by 3% of the population or by 14 million people. In addition, among EU states, the Italian language is most likely to be learned as a second language in Malta by 61% of the population, as well as in Slovenia by 15% of the population, in Croatia by 14% of the population, Austria by 11% of the population, Romania by 8% of the population, and by France and Greece by 6% of the population.[1] Italian is also one of the national languages of Switzerland, which is not a part of the European Union.[42] The Italian language is well known and studied in Albania, another non-EU member, due to its historical and geographical proximity with Italy.

Influence and derived languages

From the late nineteenth to the mid-twentieth century, thousands of Italians settled in Argentina, Uruguay, southern Brazil, and Venezuela, where they formed a strong physical and cultural presence.

In some cases, colonies were established where variants of regional (i.e. non-central) Italian languages were used, and some continue to use a derived dialect. Examples are Rio Grande do Sul, Brazil, where Talian is used, and the town of Chipilo near Puebla, Mexico; each continues to use a derived form of Venetian dating back to the nineteenth century. Another example is Cocoliche, an Italian-Spanish pidgin once spoken in Argentina and especially in Buenos Aires, and Lunfardo.

Rioplatense Spanish, and particularly the speech of the city of Buenos Aires, has intonation patterns that resemble those of Italian dialects, because Argentina has had a continuous large influx of Italian settlers since the second half of the nineteenth century: initially primarily from northern Italy; then, since the beginning of the twentieth century, mostly from southern Italy.

Lingua franca

Starting in late medieval times, Italian language variants replaced Latin to become the primary commercial language in much of Europe and the Mediterranean Sea (especially the Tuscan and Venetian variants). These variants were consolidated during the Renaissance with the strength of Italian and the rise of humanism in the arts.

During the Renaissance, Italy held artistic sway over the rest of Europe. All educated European gentlemen were expected to make the Grand Tour, visiting Italy to see its great historical monuments and works of art. It thus became expected that educated Europeans should learn at least some Italian; the English poet John Milton, for instance, wrote some of his early poetry in Italian. In England, Italian became the second most common modern language to be learned, after French (though the classical languages, Latin and Greek, came first). However, by the late eighteenth century, Italian tended to be replaced by German as the second modern language in the curriculum. Yet Italian loanwords continue to be used in most other European languages in matters of art and music.

Within the Catholic church, Italian is known by a large part of the ecclesiastical hierarchy, and is used in substitution for Latin in some official documents. The presence of Italian as the primary language in the Vatican City indicates use, not only within the Holy See, but also throughout the world where an episcopal seat is present. It continues to be used in music and opera. Other examples where Italian is sometimes used as a means of communication are in some sports (sometimes in football and motorsports) and in the design and fashion industries.

Italian dialects

In Italy, almost all Romance languages spoken as the vernacular (other than standard Italian and other unrelated, non-Italian languages) are termed "Italian dialects"; the only exceptions are Sardinian and Friulan, which the law recognises as official regional languages.

Many Italian dialects may be considered historical languages in their own right.[43] These include recognized language groups such as, Neapolitan, Sardinian, Sicilian, Ligurian, Piedmontese, Venetian, and others, and regional variants of these languages such as Calabrian. The distinction between dialect and language has been made by scholars (such as Francesco Bruni): on the one hand are the languages that made up the Italian koine; and on the other, those that had little or no part in it, such as Albanian, Greek, German, Ladin, and Occitan, which some minorities still speak. The Corsican language is also related to Italian.

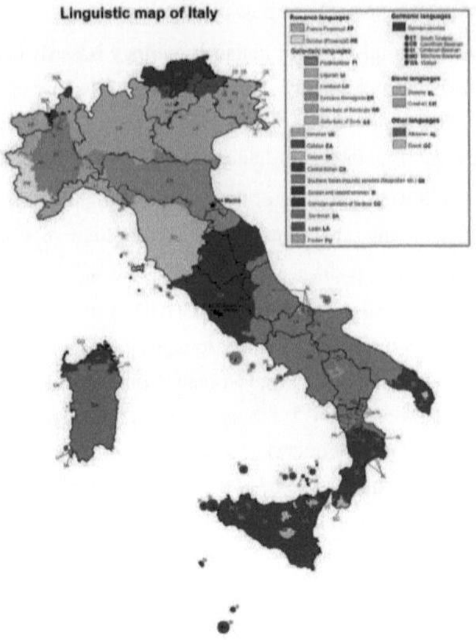

Italian languages

Regional differences can be recognized by various factors: the openness of vowels, the length of the consonants, and influence of the local dialect (for example, in informal situations the contraction *annà* replaces *andare* in the area of Rome for the infinitive "to go"; and *nare* is what Venetians say for the infinitive "to go").

Phonology

Consonants of Italian[44]

	Bilabial	Labio-dental	Dental/ Alveolar	Post-alveolar	Palatal	Velar
Nasal	m		n		ɲ	
Plosive	p b		t d			k g
Affricate			ts dz	tʃ dʒ		
Fricative		f v	s z	ʃ		
Trill			r			
Lateral			l		ʎ	
Approximant					j	w

Italian has a typical Romance-language seven-vowel system, consisting of /a, ɛ, e, i, ɔ, o u/, as well as 23 consonants. Compared with most other Romance languages, Italian phonology is extremely conservative, preserving many words nearly unchanged from Vulgar Latin. Some examples:

- Italian *quattordici* "fourteen" < Latin quattuordecim (cf. Spanish *catorce*, French *quatorze* /ka'tɔʁz/)
- Italian *settimana* "week" < Latin septimāna (cf. Spanish *semana*, French *semaine* /s(ə)'mɛn/)
- Italian *medesimo* "same" < Vulgar Latin *medi(p)simum (cf. Spanish *mismo*, French *même* /mɛm/)
- Italian *guadagnare* "to win, earn" < Vulgar Latin *guadanyāre < Germanic /waidanjan/ (cf. Catalan *guanyar*, Spanish *ganar*, French *gagner* /ga'ɲe/)

The conservativeness of Italian phonology is partly explained by its origin. Italian stems from a literary language that is derived from the 13th-century speech of the city of Florence in the region of Tuscany, and has changed little in the last 700 years or so. Furthermore, the Tuscan dialect is the most conservative of all Italian dialects, radically different from the Gallo-Italian languages less than 100 miles to the north (across the La Spezia-Rimini line).

The following are some of the conservative phonological features of Italian, as compared with the common Western Romance languages (French, Spanish, Portuguese, Catalan). Some of these features are also present in Romanian.

- Little or no lenition of consonants between vowels, e.g. vīta > *vita* "life" (cf. Spanish *vida* [biða], French *vie*), pedem > *piede* "foot" (cf. Spanish *pie*, French *pied* /pje/).
- Preservation of doubled consonants, e.g. annum > *anno* "year" (cf. Spanish *año* /aɲo/, French *an* /ɑ̃/).
- Preservation of all Proto-Romance final vowels, e.g. pacem > *pace* "peace" (cf. Spanish *paz*, French *paix* /pɛ/), octō > *otto* "eight" (cf. Spanish *ocho*, French *huit*), fēcī > *feci* "I did" (cf. Spanish *hice*, French *fis* /fi/).
- Preservation of intertonic vowels (those between the stressed syllable and either the beginning or ending syllable). This accounts for some of the most noticeable differences, as in the forms *quattordici* and *settimana* given above.
- Lack of various consonant "deformations", e.g. folia > Italo-Western /fɔʎʎa/ > *foglia* /fɔʎʎa/ "leaf" (cf. Spanish *hoja* /oxa/, French *feuille* /fœj/; but note Portuguese *folha* /foʎɐ/).

Compared with most other Romance languages, Italian has a large number of inconsistent outcomes, where the same underlying sound produces different results in different words, e.g. laxāre > *lasciare* and *lassare*, captiāre > *cacciare* and *cazzare*, (ex)dēroteolāre > *sdrucciolare* and *druzzolare*, rēgīna > *regina* and *reina*, -c- > /k/ and /g/, -t- > /t/ and /d/. This is thought to reflect the several-hundred-year period during which Italian developed as a literary language divorced from any native-speaking population, with an origin in 12th/13th-century Tuscan but with many words borrowed from dialects farther to the north, with different sound outcomes. (The La Spezia-Rimini line, the most important isogloss in the entire Romance-language area, passes only about 20 miles to the north of Florence.)

Some other features that distinguish Italian from the Western Romance languages:

- Latin ce-,ci- becomes /tʃe,tʃi/ rather than /(t)se,(t)si/.
- Latin -ct- becomes /tt/ rather than /jt/ or /tʃ/: octō > *otto* "eight" (cf. Spanish *ocho*, French *huit*).
- Vulgar Latin -cl- becomes *cchi* /kkj/ rather than /ʎ/: oclum > *occhio* "eye" (cf. Portuguese *olho* /oʎu/, French *oeil* /œj/ < /œʎ/).
- Final /s/ is not preserved, and vowel changes rather than /s/ are used to mark the plural: *amico, amici* "male friend(s)", *amica, amiche* "female friend(s)" (cf. Spanish *amigo(s)* "male friend(s)", *amiga(s)* "female friends"); trēs, sex > *tre, sei* "three, six" (cf. Spanish *tres, seis*).

Standard Italian also differs in some respects from most nearby Italian dialects:

- Perhaps most noticeable is the total lack of metaphony, a feature characterizing nearly every other Italian dialect.
- No simplification of original /nd/, /mb/ (which often became /nn/, /mm/ elsewhere).

Writing system

The Italian alphabet has only 21 letters. The letters j, k, w, x, y are excluded, though they appear in loanwords such as *jeans, whisky* and *taxi*. The letter x has become common in standard Italian with the prefix *extra-*, although *(e)stra-*[45] [46] is traditionally used. The letter j originated as an archaic orthographic variant of i. It appears in the first name *Jacopo* and in some Italian place-names, such as Bajardo, Bojano, Joppolo, Jerzu, Jesolo, Jesi, Ajaccio, among numerous others. It also appears in *Mar Jonio*, an alternative spelling of *Mar Ionio* (the Ionian Sea). The letter j may appear in dialectal words, but its use is discouraged in contemporary standard Italian. The foreign letters can be substituted with phonetically equivalent native Italian letters and digraphs: gi or ge for j; c or ch for k (including in the standard prefix *kilo-*); o, u or v for w; s, ss, z, zz or cs for x; and e or i for y.

- The acute accent is used over e to indicate a front close-mid vowel, as in *perché* "why, because". The grave accent is used over e to indicate a front open-mid vowel, as in *tè* "tea". The grave accent is used over any vowel to indicate word-final stress, as in *gioventù* "youth". The penultimate syllable is typically stressed. If the stressed vowel is the final letter of the word, the accent is mandatory, otherwise it is not mandatory (unlike in Spanish or in Greek) and virtually always omitted. When a word is potentially ambiguous, the accent is sometimes used for disambiguation, as for *prìncipi* "princes" and *princìpi* "principles". For monosyllabic words, the rule is different: when two identical monosyllabic words with different meanings exist, the accent is compulsory on one and forbidden on the other (example: *è* "is", *e* "and"). Rare, polysyllabic words can have doubtful stress. *Istanbul* can be accented on the first (*Ìstanbul*) or second syllable (*Istànbul*). The U.S. state name *Florida* is pronounced in Italian as in Spanish with stress on the second syllable (*Florìda*). Because of an Italian word with the same spelling but different stress (*flòrida* "flourishing") and because of the English pronunciation, most Italians pronounce *Florida* with stress on the first syllable. Dictionaries give the latter as an alternative pronunciation.[47]
- The letter h distinguishes *ho, hai, ha, hanno* (present indicative of *avere* "to have") from *o* ("or"), *ai* ("to the"), *a* ("to"), *anno* ("year"). In the spoken language, the letter is always silent. The h in *ho* additionally marks the contrasting open pronunciation of the o. The letter h is also used in combinations with other letters. No phoneme [h] exists in Italian. In nativised foreign words, the h is silent. For example, *hotel* and *hovercraft* are pronounced /oˈtɛl/ and /ˈɔverkraft/ respectively.
- The letters s and z can symbolize voiced or voiceless consonants. z symbolizes /dz/ or /ts/ depending on context, with few minimal pairs. For example: *zanzara* /dzanˈdzaːra/ "mosquito" and *nazione* /natˈtsjoːne/ "nation". s symbolizes /s/ word-initially before a vowel, when clustered with a voiceless consonant (p, f, c, ch), and when doubled; it symbolizes /z/ when between vowels and when clustered with voiced consonants. Intervocalic s varies regionally between /s/ and /z/.
- The letters c and g vary in pronunciation between plosives and affricates depending on following vowels. The letter c symbolizes /k/ when word-final and before the back vowels a, o, u. It symbolizes /tʃ/ as in *chair* before the front vowels e, i. The letter g symbolizes /ɡ/ when word-final and before the back vowels a, o,

u. It symbolizes /dʒ/ as in *gem* before the front vowels e, i. French, Spanish, Romanian and, to a lesser extent, English have similar variations for c, g. Compare hard and soft C, hard and soft G. (See also palatalization.)

- The digraphs ch and gh indicate or preserve hardness (/k/ and /ɡ/) before i, e. The digraphs ci and gi indicate or preserve softness (/tʃ/ and /dʒ/) before a, o, u. For example:

	Before back vowel (A, O, U)		Before front vowel (I, E)	
Plosive	C	caramella /karaˈmɛlla/ *candy*	CH	china /ˈkina/ *India ink*
	G	gallo /ˈɡallo/ *rooster*	GH	ghiro /ˈɡiro/ *edible dormouse*
Affricate	CI	ciaramella /tʃaraˈmɛlla/ *shawm*	C	Cina /ˈtʃina/ *China*
	GI	giallo /ˈdʒallo/ *yellow*	G	giro /ˈdʒiro/ *round, tour*

Note: h is silent in the digraphs *ch*, *gh*; and i is silent in the digraphs ci and gi before a, o, u unless the i is stressed. For example, it is silent in *ciao* /ˈtʃa.o/ and cielo /ˈtʃɛ.lo/, but it is pronounced in *farmacia* /ˌfar.maˈtʃi.a/ and *farmacie* /ˌfar.maˈtʃi.e/.

Italian has geminate, or double, consonants, which are distinguished by length. Length is distinctive for all consonants except for /ʃ/, /ts/, /dz/, /ʎ/, /ɲ/, which are always geminate, and /z/, which is always single. Geminate plosives and affricates are realised as lengthened closures. Geminate fricatives, nasals, and /l/ are realized as lengthened continuants. There is only one vibrant phoneme /r/ but the actual pronunciation depends on context and regional accent. Generally one can find a flap consonant [ɾ] in unstressed position while [r] is more common in stressed syllables, but there may be exceptions. Especially people from the Northern part of Italy (Parma, Aosta Valley, South Tyrol) may pronounce /r/ as [ʀ], [ʁ], or [ʋ].[48]

Of special interest to the linguistic study of Italian is the *gorgia toscana*, or "Tuscan Throat", the weakening or lenition of certain intervocalic consonants in Tuscan dialects.

The voiced postalveolar fricative /ʒ/ is only present in loanwords: for example, *garage* [ɡaˈraːʒ].

Assimilation

Italian phonotactics do not usually permit verbs and polysyllabic nouns to end with consonants, excepting poetry and song, so foreign words may receive extra terminal vowel sounds.

Grammar

Italian grammar is typical of the grammar of Romance languages in general. Cases exist for pronouns (nominative, oblique, accusative, dative), but not for nouns. There are two genders (masculine and feminine). Nouns, adjectives, and articles inflect for gender and number (singular and plural). Adjectives are sometimes placed before their noun and sometimes after. Subject nouns generally come before the verb. Subjective pronouns are usually dropped, their presence implied by verbal inflections. Noun objects come after the verb, as do pronoun objects after imperative verbs and infinitives, but otherwise pronoun objects come before the verb. There are numerous contractions of prepositions with subsequent articles. There are numerous productive suffixes for diminutive, augmentative, pejorative, attenuating etc., which are also used to create neologisms.

There are three regular sets of verbal conjugations, and various verbs are irregularly conjugated. Within each of these sets of conjugations, there are four simple (one-word) verbal conjugations by person/number in the indicative mood (present tense; past tense with imperfective aspect, past tense with perfective aspect, and future tense), two simple conjugations in the subjunctive mood (present tense and past tense), one simple conjugation in the conditional mood, and one simple conjugation in the imperative mood. Corresponding to each of the simple conjugations, there is a compound conjugation involving a simple conjugation of "to be" or "to have" followed by a past participle.

Examples

Numbers

English	Italian	IPA
One	*uno*	/ˈuno/
Two	*due*	/ˈdue/
Three	*tre*	/tre/
Four	*quattro*	/ˈkwattro/
Five	*cinque*	/ˈtʃiŋkwe/
Six	*sei*	/ˈsɛi/
Seven	*sette*	/ˈsɛtte/
Eight	*otto*	/ˈɔtto/
Nine	*nove*	/ˈnɔve/
Ten	*dieci*	/ˈdjɛtʃi/

English	Italian	IPA
Eleven	*undici*	/ˈunditʃi/
Twelve	*dodici*	/ˈdoditʃi/
Thirteen	*tredici*	/ˈtreditʃi/
Fourteen	*quattordici*	/kwatˈtorditʃi/
Fifteen	*quindici*	/ˈkwinditʃi/
Sixteen	*sedici*	/ˈsɛditʃi/
Seventeen	*diciassette*	/ditʃasˈsɛtte/
Eighteen	*diciotto*	/diˈtʃɔtto/
Nineteen	*diciannove*	/ditʃanˈnɔve/
Twenty	*venti*	/ˈventi/

English	Italian	IPA
Twenty-one	*ventuno*	/venˈtuno/
Twenty-two	*ventidue*	/ventiˈdue/
Twenty-three	*ventitre*	/ventiˈtre/
Twenty-four	*ventiquattro*	/ventiˈkwattro/
Twenty-five	*venticinque*	/ventiˈtʃinkwe/
Twenty-six	*ventisei*	/ventiˈsɛi/
Twenty-seven	*ventisette*	/ventiˈsɛtte/
Twenty-eight	*ventotto*	/venˈtɔtto/
Twenty-nine	*ventinove*	/ventiˈnɔve/
Thirty	*trenta*	/ˈtrenta/

English	Italian
one hundred	cento
one thousand	mìlle
two thousand	duemila
two thousand twelve {2012}	duemiladodici

Days of the week

English	Italian	IPA
Monday	*lunedì*	/lune'di/
Tuesday	*martedì*	/marte'di/
Wednesday	*mercoledì*	/merkole'di/
Thursday	*giovedì*	/dʒove'di/
Friday	*venerdì*	/vener'di/
Saturday	*sabato*	/'sabato/
Sunday	*domenica*	/do'menika/

Sample texts

There is a recording of Dante's Divine Comedy read by Lino Pertile available at http://etcweb.princeton.edu/dante/pdp/

See also

- Accademia della Crusca
- CELI
- CILS (Qualification)
- Enciclopedia Italiana
- Guide to phonetic transliteration of Italian
- Italian alphabet
- Italian dialects
- Italian exonyms
- Italian grammar
- Italian honorifics
- The Italian Language Foundation (in the United States)
- Italian language in Slovenia
- Italian literature
- Italian musical terms
- Italian phonology
- Italian profanity
- Italian Sign Language
- Italian Studies
- Italian Wikipedia
- Italian-language international radio stations

- List of languages of Italy
- Sicilian School
- Veronese Riddle
- Languages of the Vatican City

References and notes

[1] Eurobarometer – Europeans and their languages (http://ec.europa.eu/public_opinion/archives/ebs/ebs_243_sum_en.pdf)PDF (485 KB), February 2006

[2] Ethnologue report for language code:ita (Italy) (http://www.ethnologue.com/show_language.asp?code=ita) – Gordon, Raymond G., Jr. (ed.), 2005. Ethnologue: Languages of the World, Fifteenth edition. Dallas, Tex.: SIL International. Online version

[3] Languages of Italy (http://www.ethnologue.com/show_country.asp?name=IT) - Ethnologue - Languages of the World - Copyright © 2010 SIL International.

[4] *Legge sulle fonti del diritto* of 7 June 1929, laws and regulations are published in the Italian-language *Supplemento per le leggi e disposizioni dello Stato della Città del Vaticano* attached to the [[Acta Apostolicae Sedis (http://www.vaticanstate.va/NR/rdonlyres/FBFEA0E8-B43A-452A-AAA0-1AF49590F658/2615/Supplemento.pdf)].]

[5] Modern Italian (http://www.italian-language-study.com/italian-language/modern-italian.htm) *The Italian Language* Retrieved 2010-05-16

[6] Grimes, Barbara F. (October 1996). Barbara F. Grimes. ed. *Ethnologue: Languages of the World*. Consulting Editors: Richard S. Pittman & Joseph E. Grimes (thirteenth ed.). Dallas, Texas: Summer Institute of Linguistics, Academic Pub. ISBN 1-55671-026-7.

[7] "History of the Italian language." (http://www.italian-language.biz/italian/history.asp). . Retrieved 2006-09-24.

[8] "Lewis, M. Paul (ed.), 2009. Ethnologue: Languages of the World, Sixteenth edition" (http://www.ethnologue.com/show_language.asp?code=ita). Ethnologue.com. . Retrieved 2010-04-21.

[9] Brincat (2005)

[10] "Languages Spoken by More Than 10 Million People" (http://www.webcitation.org/query?id=1257013011437361). *Languages Spoken by More Than 10 Million People*. Microsoft Encarta 2006. Archived from the original (http://encarta.msn.com/media_701500404/Languages_Spoken_by_More_Than_10_Million_People.html) on 2009-10-31. . Retrieved 2007-02-18.

[11] Associazioneitalianadikerc (http://www.sovietmemories.it/schede/Associazioneitalianadikerc.htm)

[12] "1,500,000 mother tongue Italian speakers in Argentina" (http://www.ethnologue.com/show_country.asp?name=AR). Ethnologue.com. . Retrieved 2010-04-21.

[13] American FactFinder, United States Census Bureau. "over 1 million Americans speak Italian at home" (http://factfinder.census.gov/servlet/IPTable?_bm=y&-reg=ACS_2005_EST_G00_S0201:543;ACS_2005_EST_G00_S0201PR:543;ACS_2005_EST_G00_S0201T:543;ACS_2005_EST_G00_S0201TPR:543&-qr_name=ACS_2005_EST_G00_S0201&-qr_name=ACS_2005_EST_G00_S0201PR&-qr_name=ACS_2005_EST_G00_S0201T&-qr_name=ACS_2005_EST_G00_S0201TPR&-ds_name=ACS_2005_EST_G00_&-TABLE_NAMEX=&-ci_type=A&-redoLog=false&-charIterations=031&-geo_id=01000US&-format=&-_lang=en). Factfinder.census.gov. . Retrieved 2010-04-21.

[14] "Statistics Canada 2006" (http://www12.statcan.ca/english/census06/data/topics/RetrieveProductTable.cfm?ALEVEL=3&APATH=3&CATNO=&DETAIL=0&DIM=&DS=99&FL=0&FREE=0&GAL=0&GC=99&GK=NA&GRP=1&IPS=&METH=0&ORDER=1&PID=89189&PTYPE=88971&RL=0&S=1&ShowAll=No&StartRow=1&SUB=705&Temporal=2006&Theme=70&VID=0&VNAMEE=&VNAMEF=&GID=837928). 2.statcan.ca. 2010-04-08. . Retrieved 2010-04-21.

[15] "548,000 mother tongue Italian speakers in Germany" (http://www.ethnologue.com/show_country.asp?name=DE). Ethnologue.com. . Retrieved 2010-04-21.

[16] Vannini, Marisa. *Italia y los Italianos en la Historia y en la Cultura de Venezuela*. Oficina Central de Información (Ministerio del Interior). Caracas, 1966

[17] "353,605 mother tongue Italian speakers in Australia" (http://www.ethnologue.com/show_country.asp?name=AU). Ethnologue.com. . Retrieved 2010-04-21.

[18] "200,000 mother tongue Italian speakers in the UK" (http://www.ethnologue.com/show_country.asp?name=GB). Ethnologue.com. . Retrieved 2010-04-21.

[19] "79,000 mother tongue Italian speakers in Uruguay" (http://www.ethnologue.com/show_country.asp?name=Uy). Ethnologue.com. . Retrieved 2010-04-21.

[20] "72,400 mother tongue Italian speakers in Egypt" (http://www.ethnologue.com/show_country.asp?name=EG). Ethnologue.com. . Retrieved 2010-04-21.

[21] "50,000 mother tongue Italian speakers in Brazil" (http://www.ethnologue.com/show_country.asp?name=BR). Ethnologue.com. . Retrieved 2010-04-21.

[22] "2,25o mother tongue Italian speakers in Colombia" (http://www.ethnologue.com/show_country.asp?name=AR). Ethnologue.com. . Retrieved 2010-04-21.

[23] "Ethnologue report for France" (http://www.ethnologue.com/show_country.asp?name=FR). Ethnologue.com. . Retrieved 2010-04-21.

[24] Languages of Eritrea – Tigrinya (http://home.planet.nl/~hans.mebrat/eritrea-languages.htm)

[25] "Scuola Italiana di Asmara (in Italian)" (http://www.scuoleasmara.it). Scuoleasmara.it. . Retrieved 2010-04-21.
[26] Tekle M. Woldemikael, "Language, Education, and Public Policy in Eritrea," in African Studies Review, Vol. 46, No. 1. (Apr., 2003), pp. 117–136.
[27] "Eritrea" (http://lcweb2.loc.gov/frd/cs/profiles/Eritrea.pdf) (PDF). . Retrieved 2010-04-21.
[28] Boston, Massachusetts (http://www.mla.org/map_data_results&state_id=25&place_id=7000&cty_id=), *MLA Data Center*
[29] Chicago, Illinois (http://www.mla.org/map_data_results&state_id=17&place_id=14000&cty_id=), *MLA Data Center*
[30] Data Center Results (http://www.mla.org/cgi-shl/docstudio/docs.pl?map_data_results)
[31] New York, New York (http://www.mla.org/map_data_results&state_id=36&place_id=51000&cty_id=), *MLA Data Center*
[32] Philadelphia, Pennsylvania (http://www.mla.org/map_data_results&state_id=42&place_id=60000&cty_id=), *MLA Data Center*
[33] "Table 5. Detailed List of Languages Spoken at Home for the Population 5 Years and Over by State: 2000" (http://www.census.gov/population/cen2000/phc-t20/tab05.pdf) (PDF). . Retrieved 2010-04-21.
[34] Australian Bureau of Statistics, 2005, "Language other than English" (spreadsheet of figures from 2001 Census) (http://abs.gov.au/websitedbs/D3310116.NSF/85255e31005a1918852556c2005508d8/c47ad86d67c1466bca256ce0007e8d6b/$FILE/ATTH23CO/Exstatic 2 2005, Australia.xls)
[35] Australian Bureau of Statistics, 2002, "A Snapshot of Melbourne" (http://www8.abs.gov.au/censusoutput/abs@CPP.nsf/Lookup/205Snapshot12001?OpenDocument&TabName=Summary&ProdNo=205&Issue=2001&Num=&View=&)
[36] Australian Bureau of Statistics, 2002, "A Snapshot of Sydney" (http://www8.abs.gov.au/censusoutput/abs@CPP.nsf/Lookup/105Snapshot12001#Ancestry)
[37] "9" (http://www.iic-colonia.de/italiano-2000/09.12 Analisi generale dei dati.htm). Iic-colonia.de. . Retrieved 2010-04-21.
[38] Dati del Ministero degli Affari Esteri (http://www.esteri.it/MAE/IT/Politica_Estera/Cultura/PromozioneLinguaItaliana/DatiStatisticheInsegnamentoLingua.htm)
[39] Language-learning trends in the United States (http://www.vistawide.com/languages/us_languages.htm) *VistaWide* Retrieved 2010-05-16
[40] http://www.repubblica.it/2007/04/sezioni/esteri/italiano-rivincita/italiano-rivincita/italiano-rivincita.html?ref=search
[41] "www.iic-colonia.de" (http://www.iic-colonia.de/italiano-2000/Indice.htm). www.iic-colonia.de. . Retrieved 2010-04-21.
[42] A language of Italy (http://www.ethnologue.com/show_language.asp?code=ita) *Ethnologue* Retrieved 2010-06-05
[43] "Ethnologue web reference for Italian" (http://www.ethnologue.com/14/show_language.asp?code=ITN). Ethnologue.com. . Retrieved 2010-04-21.
[44] Rogers & d'Arcangeli (2004:117)
[45] http://translate.google.com/#en|it|extrapolate
[46] http://translate.google.com/#en|it|extraordinary
[47] **(Italian)** Dizionario d'ortografia e di pronunzia (http://www.dizionario.rai.it/poplemma.aspx?lid=56814&r=1715)
[48] Canepari, Luciano (January 1999). *Il MªPI – Manuale di pronuncia italiana* (second ed.). Bologna: Zanichelli. ISBN 88-08-24624-8.

Bibliography

- Rogers, Derek; d'Arcangeli, Luciana (2004). "Italian". *Journal of the International Phonetic Association* **34** (1): 117–121. doi:10.1017/S0025100304001628
- M. Vitale, *Studi di Storia della Lingua Italiana*, LED Edizioni Universitarie, Milano, 1992, ISBN 88-7916-015-X
- S. Morgana, *Capitoli di Storia Linguistica Italiana*, LED Edizioni Universitarie, Milano, 2003, ISBN 88-7916-211-X
- J. Kinder, *CLIC: Cultura e Lingua d'Italia in CD-ROM / Culture and Language of Italy on CD-ROM*, Interlinea, Novara, 2008, ISBN 978-88-8212-637-7

External links

- Swadesh list in English and Italian
- Italian proverbs
- The online edition (2007) of the *Dizionario d'ortografia e di pronunzia* (*DOP*), a pronouncing dictionary of standard Italian (http://www.dizionario.rai.it), *RAI*
- " Learn Italian (http://www.bbc.co.uk/languages/italian/)," *BBC*
- USA Foreign Service Institute Italian basic course (http://www.fsi-language-courses.org/Content.php?page=Italian)

Venice

Venice
Venezia

— Comune —

Comune di Venezia

A collage of Venice: at the top left is the Piazza San Marco, followed by a view of the city, then the Grand Canal, and (smaller) the interior of La Fenice and finally the Island of San Giorgio Maggiore

Coat of arms

Venice

Location of Venice in Italy

Coordinates: 45°26′15″N 12°20′9″E

Country	Italy
Region	Veneto
Province	Venice (VE)
Frazioni	Chirignago, Favaro Veneto, Mestre, Marghera, Murano, Burano, Giudecca, Lido, Zelarino

Government	
• Mayor	Giorgio Orsoni (Democratic Party)
Area	
• Total	414.57 km^2 (**unknown operator: u'strong'** sq mi)
Elevation	0 m (0 ft)
Population (2009-04-30)	
• Total	270660
• Density	**unknown operator: u'strong'**/km^2 (**unknown operator: u'strong'**/sq mi)
Demonym	Venetians
Time zone	CET (UTC+1)
• Summer (DST)	CEST (UTC+2)
Postal code	30100
Dialing code	041
Patron saint	St. Mark the Evangelist
Saint day	25 April
Website	Official website [1]

Venice and its Lagoon *

UNESCO World Heritage Site

Country	Italy
Type	Cultural
Criteria	i, ii, iii, iv, v, vi
Reference	394 [2]
Region **	Europe and North America
Inscription history	
Inscription	1987 (11th Session)

* Name as inscribed on World Heritage List [3]
** Region as classified by UNESCO [4]

Venice (Italian: *Venezia* Italian pronunciation: [veˈnɛttsja] (listen), Venetian: *Venexia* [veˈnɛsja]) is a city in northeast Italy sited on a group of 118 small islands separated by canals and linked by bridges.[5] It is located in the marshy Venetian Lagoon which stretches along the shoreline between the mouths of the Po and the Piave Rivers. Venice is renowned for the beauty of its setting, its architecture and its artworks.[5] The city and lagoon are listed as World Heritage Sites.[5]

Venice is the capital of the Veneto region. In 2009, there were 270,098 people residing in Venice's comune (the population estimate of 272,000 inhabitants includes the population of the whole Comune of Venezia; around 60,000[6] in the historic city of Venice (*Centro storico*); 176,000 in *Terraferma* (the *Mainland*), mostly in the large *frazioni* of Mestre and Marghera; 31,000 live on other islands in the lagoon). Together with Padua and Treviso, the city is included in the Padua-Treviso-Venice Metropolitan Area (PATREVE) (population 1,600,000).

The name is derived from the ancient Veneti people who inhabited the region by the 10th century B.C.[7] [8] The city historically was the capital of the Venetian Republic. Venice has been known as the "La Dominante", "Serenissima", "Queen of the Adriatic", "City of Water", "City of Masks", "City of Bridges", "The Floating City", and "City of Canals". Luigi Barzini described it in *The New York Times* as "undoubtedly the most beautiful city built by man".[9] Venice has also been described by the *Times Online* as being one of Europe's most romantic cities.[10]

The Republic of Venice was a major maritime power during the Middle Ages and Renaissance, and a staging area for the Crusades and the Battle of Lepanto, as well as a very important center of commerce (especially silk, grain, and spice) and art in the 13th century up to the end of the 17th century. This made Venice a wealthy city throughout most of its history.[11] It is also known for its several important artistic movements, especially the Renaissance period. Venice has played an important role in the history of symphonic and operatic music, and it is the birthplace of Antonio Vivaldi.

History

Origins

While there are no historical records that deal directly with the obscure and peripheral origins of Venice,[12] tradition and the available evidence have led several historians to agree that the original population of Venice consisted of refugees from Roman cities near Venice such as Padua, Aquileia, Treviso, Altino and Concordia (modern Portogruaro) and from the undefended countryside, who were fleeing successive waves of Germanic and Hun invasions.[13] Some late Roman sources reveal the existence of fishermen on the islands in the original marshy lagoons. They were referred to as *incolae lacunae* ("lagoon dwellers"). The traditional founding is identified with the dedication of the first church, that of San Giacomo at the islet of Rialto (Rivoalto, "High Shore"), which is said to have been at the stroke of noon on 25 March 421.[14] [15]

The last and most enduring immigration into the north of the Italian peninsula was that of the Lombards in 568, leaving the Eastern Roman Empire a small strip of coast in the current Veneto, and the main administrative and religious entities were therefore transferred to this remaining dominion, centered upon the Exarchate of Ravenna, the local representative of the Emperor in the East. The Venetian tradition of the islanders' aid to Belisarius was reported in early histories to explain the largely theoretical link to Ravenna, and to the Eastern Emperor. New ports were built, including those at Malamocco and Torcello in the Venetian lagoon. The *tribuni maiores*, the earliest central standing governing committee of the islands in the Lagoon, dated from c. 568.[16]

The Venetians offered asylum to the Exarch Paul, who was in flight from the Lombard Liutprand.[17] Byzantine domination of central and northern Italy was subsequently largely eliminated by the conquest of the Exarchate of Ravenna in 751 by Aistulf. During this period, the seat of the local Byzantine governor (the "duke/dux", later "doge") was situated in Malamocco. Settlement on the islands in the lagoon probably increased in correspondence with the Lombard conquest of the Byzantine territories.

Sometime in the first decades of the eighth century, the people of the lagoon elected their first leader Ursus, who was confirmed by Byzantium and given the titles of *hypatus* and *dux*.[18] He was the first historical Doge of Venice.

In 775/776, the episcopal seat of Olivolo (Helipolis) was created. During the reign of duke Agnello Particiaco (811–827) the ducal seat was moved from Malamocco to the highly protected Rialto, the current location of Venice. The monastery of St. Zachary and the first ducal palace and basilica of St. Mark, as well as a walled defense (civitatis murus) between Olivolo and Rialto were subsequently built here. Winged lions, which may be seen throughout Venice, are a symbol for St. Mark.

Charlemagne was initially hostile to Venice and sought to subdue the city to his own rule. He ordered the Pope to expel the Venetians from the Pentapolis along the Adriatic coast,[19] and Charlemagne's own son Pepin of Italy, king of the Lombards under the authority of his father, embarked on a siege of Venice itself. This, however, proved a costly failure. The siege lasted six months, with Pepin's army ravaged by the diseases of the local swamps and eventually forced to withdraw. A few months later Pepin himself died, apparently as a result of a disease contracted there. In the aftermath, an agreement between Charlemagne and Nicephorus recognized Venice as Byzantine territory and recognized the city's trading rights along the Adriatic coast.

In 828, the new city's prestige was raised by the acquisition of the claimed relics of St. Mark the Evangelist from Alexandria, which were placed in the new basilica. The patriarchal seat was also moved to Rialto. As the community continued to develop and as Byzantine power waned, it led to the growth of autonomy and eventual independence.

Expansion

From the ninth to the twelfth century Venice developed into a city state (an Italian thalassocracy or *Repubblica Marinara*, the other three being Genoa, Pisa, and Amalfi). Its strategic position at the head of the Adriatic made Venetian naval and commercial power almost invulnerable. With the elimination of pirates along the Dalmatian coast, the city became a flourishing trade center between Western Europe and the rest of the world (especially the Byzantine Empire and the Islamic world). In the 12th century the foundations of Venice's power were laid: the Venetian Arsenal was under construction in 1104; the last autocratic doge, Vital II Michele, died in 1172.

Piazza San Marco in Venice, with St Mark's Campanile and Basilica in the background

The Republic of Venice seized a number of places on the eastern shores of the Adriatic before 1200, mostly for commercial reasons, because pirates based there were a menace to trade. The Doge already carried the titles of Duke of Dalmatia and Duke of Istria. Later mainland possessions, which extended across Lake Garda as far west as the Adda River, were known as the "Terraferma", and were acquired partly as a buffer against belligerent neighbours, partly to guarantee Alpine trade routes, and partly to ensure the supply of mainland wheat, on which the city depended. In building its maritime commercial empire, the Republic dominated the trade in salt,[20] acquired control of most of the islands in the Aegean, including Cyprus and Crete, and became a major power-broker in the Near East. By the standards of the time, Venice's stewardship of its mainland territories was relatively enlightened and the citizens of such towns as Bergamo, Brescia and Verona rallied to the defence of Venetian sovereignty when it was threatened by invaders.

These Horses of Saint Mark are a replica of the *Triumphal Quadriga* captured in Constantinople in 1204 and carried to Venice as a trophy.

Venice remained closely associated with Constantinople, being twice granted trading privileges in the Eastern Roman Empire, through the so called Golden Bulls or 'chrysobulls' in return for aiding the Eastern Empire to resist Norman and Turkish incursions. In the first chrysobull Venice acknowledged its homage to the Empire but not in the second, reflecting the decline of Byzantium and the rise of Venice's power.[21] [22]

Venice became an imperial power following the Venetian-financed Fourth Crusade, which in 1204 seized and sacked Constantinople and established the Latin Empire. As a result of this conquest considerable Byzantine plunder was brought back to Venice. This plunder included the gilt bronze horses from the Hippodrome of Constantinople, which were originally placed above the entrance to St Mark's cathedral in Venice, although the originals have been replaced with replicas and the originals are now stored within the basilica. Following the fall of Constantinople the former Roman Empire was partitioned among the Latin crusaders and the Venetians. Venice subsequently carved out a sphere of influence in the Mediterranean known as the Duchy of the Archipelago, and seized Crete.

The seizure of Constantinople would ultimately prove as decisive a factor in ending the Byzantine Empire as the loss of the Anatolian themes after Manzikert. Though the Byzantines recovered control of the ravaged city a half century later, the Byzantine Empire was terminally weakened, and existed as a ghost of its old self until Sultan Mehmet The Conqueror took the city in 1453.

View of San Giorgio Maggiore Island from St. Mark's Campanile

Situated on the Adriatic Sea, Venice always traded with the Byzantine Empire and the Muslim world extensively. By the late thirteenth century, Venice was the most prosperous city in all of Europe. At the peak of its power and wealth, it had 36,000 sailors operating 3,300 ships, dominating Mediterranean commerce. During this time, Venice's leading families vied with each other to build the grandest palaces and support the work of the greatest and most talented artists. The city was governed by the Great Council, which was made up of members of the noble families of Venice. The Great Council appointed all public officials and elected a Senate of 200 to 300 individuals. Since this group was too large for efficient administration, a Council of Ten (also called the Ducal Council or the Signoria), controlled much of the administration of the city. One member of the great council was elected "Doge", or duke, the ceremonial head of the city, who normally held the title until his death.

The Venetian governmental structure was similar in some ways to the republican system of ancient Rome, with an elected chief executive (the Doge), a senate-like assembly of nobles, and a mass of citizens with limited political power, who originally had the power to grant or withhold their approval of each newly elected Doge. Church and various private properties were tied to military service, though there was no knight tenure within the city itself. The *Cavalieri di San Marco* was the only order of chivalry ever instituted in Venice, and no citizen could accept or join a foreign order without the government's consent. Venice remained a republic throughout its independent period and politics and the military were kept separate, except when on occasion the Doge personally headed the military. War was regarded as a continuation of commerce by other means (hence, the city's early production of large numbers of mercenaries for service elsewhere, and later its reliance on foreign mercenaries when the ruling class was preoccupied with commerce).

The chief executive was the Doge, who theoretically held his elective office for life. In practice, several Doges were forced by pressure from their oligarchical peers to resign the office and retire into monastic seclusion when they were felt to have been discredited by perceived political failure.

Francesco Guardi, *The Grand Canal*, 1760 (Art Institute of Chicago)

Though the people of Venice generally remained orthodox Roman Catholics, the state of Venice was notable for its freedom from religious fanaticism and it enacted not a single execution for religious heresy during the Counter-Reformation. This apparent lack of zeal contributed to Venice's frequent conflicts with the Papacy. In this context, the writings of the Anglican Divine, William Bedell, are particularly illuminating. Venice was threatened with the interdict on a number of occasions and twice suffered its imposition. The second, most famous, occasion in 1606, by order of Pope Paul V.

Venetian ambassadors sent home still-extant secret reports of the politics and rumours of European courts, providing fascinating information to modern historians.

The newly invented German printing press spread rapidly throughout Europe in the fifteenth century, and Venice was quick to adopt it. By 1482 Venice was the printing capital of the world, and the leading printer was Aldus Manutius, who invented the concept of paperback books that could be carried in a saddlebag. His *Aldine Editions* included translations of nearly all the known Greek manuscripts of the era.[23]

Decline

Venice's long decline started in the 15th century, when it first made an unsuccessful attempt to hold Thessalonica against the Ottomans (1423–1430). It also sent ships to help defend Constantinople against the besieging Turks (1453). After Constantinople fell to Sultan Mehmet II he declared war on Venice. The war lasted thirty years and cost Venice much of her eastern Mediterranean possessions. Next, Christopher Columbus discovered the New World. Then Portugal found a sea route to India, destroying Venice's land route monopoly. France, England and Holland followed them. Venice's oared galleys were at a disadvantage when it came to traversing the great oceans, and therefore Venice was left behind in the race for colonies.

The Black Death devastated Venice in 1348 and once again between 1575 and 1577.[24] In three years the plague killed some 50,000 people.[25] In 1630, the plague killed a third of Venice's 150,000 citizens.[26] Venice began to lose its position as a center of international trade during the later part of the Renaissance as Portugal became Europe's principal intermediary in the trade with the East, striking at the very foundation of Venice's great wealth, while France and Spain fought for hegemony over Italy in the Italian Wars, marginalising its political influence. However, the Venetian empire was a major exporter of agricultural products and, until the mid-18th century, a significant manufacturing center.

Military and naval affairs

By 1303, crossbow practice had become compulsory in the city, with citizens training in groups. As weapons became more expensive and complex to operate, professional soldiers were assigned to help work merchant sailing ships and as rowers in galleys. The company of "Noble Bowmen" was recruited in the later 14th century from among the younger aristocracy and served aboard both war-galleys and as armed merchantmen, with the privilege of sharing the captain's cabin.

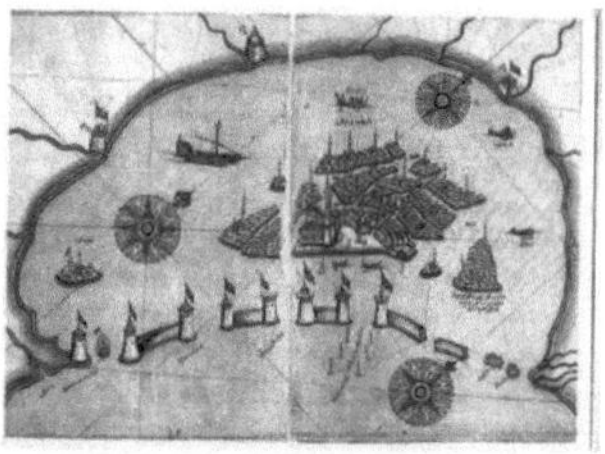

Historic map of Venice by Piri Reis

Though Venice was famous for its navy, its army was equally effective. In the 13th century, most Italian city states already were hiring mercenaries, but Venetian troops were still recruited from the lagoon, plus feudal levies from Dalmatia (the very famous *Schiavoni* or *Oltremarini*)[27] and Istria. In times of emergency, all males between seventeen and sixty years were registered and their weapons were surveyed, with those called to actually fight being organized into companies of twelve. The register of 1338 estimated that 30,000 Venetian men were capable of bearing arms; many of these were skilled crossbowmen. As in other Italian cities, aristocrats and other wealthy men were cavalrymen while the city's conscripts fought as infantry.

By 1450, more than 3,000 Venetian merchant ships were in operation. Most of these could be converted when necessary into either warships or transports. The government required each merchant ship to carry a specified number of weapons (mostly crossbows and javelins) and armour; merchant passengers were also expected to be armed and to fight when necessary. A reserve of some 25 (later 100) war-galleys was maintained in the Arsenal. Galley slaves did not exist in medieval Venice, the oarsmen coming from the city itself or from its possessions, especially Dalmatia. Those from the city were chosen by lot from each parish, their families being supported by the remainder of the parish while the rowers were away. Debtors generally worked off their obligations rowing the galleys. Rowing skills were encouraged through races and regattas.

Early in the 15th century, as new mainland territories were expanded, the first standing army was organized, consisting of *condottieri* on contract. In its alliance with Florence in 1426, Venice agreed to supply 8,000 cavalry and 3,000 infantry in time of war, and 3,000 and 1,000 in peacetime. Later in that century, uniforms were adopted that featured red-and-white stripes, and a system of honors and pensions developed. Throughout the 15th century, Venetian land forces were almost always on the offensive and were regarded as the most effective in Italy, largely because of the tradition of all classes carrying arms in defense of the city and official encouragement of general military training.

Venice, by Bolognino Zaltieri, 1565.

The command structure in the army was different from that of the fleet. By ancient law, no nobleman could command more than twenty-five men (to prevent the possibility of sedition by private armies), and while the position of Captain General was introduced in the mid-14th century, he still had to answer to a civilian panel of twenty Savi or "wise men". Not only was efficiency *not* degraded, this policy saved Venice from the military takeovers that other Italian city states so often experienced. A civilian commissioner (not unlike a commissar) accompanied each army to keep an eye on things, especially the mercenaries. The Venetian military tradition also was notably cautious; they were more interested in achieving success with a minimum expense of lives and money than in the pursuit of glory.

Modern age

The Republic lost independence when Napoleon Bonaparte conquered Venice on 12 May 1797 during the First Coalition. The French conqueror brought to an end the most fascinating century of its history: during the 18th century Venice became perhaps the most elegant and refined city in Europe, greatly influencing art, architecture and literature. Napoleon was seen as something of a liberator by the city's Jewish population, although it can be argued they had lived with fewer restrictions in Venice. He removed the gates of the Ghetto and ended the restrictions on when and where Jews could live and travel in the city.

A map of the sestiere of San Marco

Venice became Austrian territory when Napoleon signed the Treaty of Campo Formio on 12 October 1797. The Austrians took control of the city on 18 January 1798. It was taken from Austria by the Treaty of Pressburg in 1805 and became part of Napoleon's Kingdom of Italy, but was returned to Austria following Napoleon's defeat in 1814, when it became part of the Austrian-held Kingdom of Lombardy-Venetia. In 1848–1849 a revolt briefly reestablished the Venetian Republic under Daniele Manin. In 1866, following the Third Italian War of Independence, Venice, along with the rest of the Veneto, became part of the newly created Kingdom of Italy.

During the Second World War, the historic city was largely free from attack, the only aggressive effort of note being Operation Bowler, a precision strike on the German naval operations there in 1945. However the industrial areas in Mestre and Marghera and the railway lines to Padua, Trieste and Trento were repeatedly bombed.[28] On 29 April

1945 New Zealand troops under Freyberg reached Venice and relieved the city and the mainland, which were already in partisan hands.[29]

Subsidence

Further information: Acqua Alta

Acqua Alta or high water in Venice.

Venice and surroundings in false colour, from Terra. The picture is oriented with North at the top.

Foundations

The buildings of Venice are constructed on closely spaced wooden piles. Most of these piles are still intact after centuries of submersion. The foundations rest on the piles, and buildings of brick or stone sit above these footings. The piles penetrate a softer layer of sand and mud until they reach a much harder layer of compressed clay.

Submerged by water, in oxygen-poor conditions, wood does not decay as rapidly as on the surface. It is petrified as a result of the constant flow of mineral-rich water around and through it, so that it becomes a stone-like structure.

Most of these piles were made from trunks of alder trees,[30] a wood noted for its water resistance.[31] The alder came from the western-most part of today's Slovenia (resulting in the barren land of the Kras region), in two regions of Croatia, Lika and Gorski kotar (resulting in the barren slopes of Velebit) and south of Montenegro. Leonid Grigoriev has stated that Russian larch was imported to build some of Venice's foundations.[32] Larch is also used in the production of Venice turpentine.[33]

History

The city is often threatened by flood tides pushing in from the Adriatic between autumn and early spring. Six hundred years ago, Venetians protected themselves from land-based attacks by diverting all the major rivers flowing into the lagoon and thus preventing sediment from filling the area around the city. This created an ever-deeper lagoon environment.

In 1604, to defray the cost of flood relief Venice introduced what could be considered the first example of a 'stamp tax'. When the revenue fell short of expectations in 1608, Venice introduced paper with the superscription 'AQ' and imprinted instructions, which was to be used for 'letters to officials'. At first, this was to be a temporary tax, but it remained in effect until the fall of the Republic in 1797. Shortly after the introduction of the tax, Spain produced similar paper for general taxation purposes, and the practice spread to other countries.

During the 20th century, when many artesian wells were sunk into the periphery of the lagoon to draw water for local industry, Venice began to subside. It was realised that extraction of water from the aquifer was the cause. The sinking has slowed markedly since artesian wells were banned in the 1960s. However, the city is still threatened by more frequent low-level floods (called *Acqua alta*, "high water") that creep to a height of several centimetres over its quays, regularly following certain tides. In many old houses the former staircases used to unload goods are now flooded, rendering the former ground floor uninhabitable.

Some recent studies have suggested that the city is no longer sinking,[34] [35] but this is not yet certain; therefore, a state of alert has not been revoked. In May 2003 the Italian Prime Minister Silvio Berlusconi inaugurated the MOSE project (Modulo Sperimentale Elettromeccanico), an experimental model for evaluating the performance of hollow floatable gates; the idea is to fix a series of 78 hollow pontoons to the sea bed across the three entrances to the lagoon. When tides are predicted to rise above 110 centimetres, the pontoons will be filled with air, causing them to float and block the incoming water from the Adriatic Sea. This engineering work is due to be completed by 2014.[36]

Panorama of the Giudecca Canal and the Saint Mark's Basin

Geography

The city is divided into six areas or "sestiere". These are Cannaregio, San Polo, Dorsoduro (including the Giudecca and Isola Sacca Fisola), Santa Croce, San Marco (including San Giorgio Maggiore) and Castello (including San Pietro di Castello and Sant'Elena). Each sestiere was administered by a procurator and his staff.

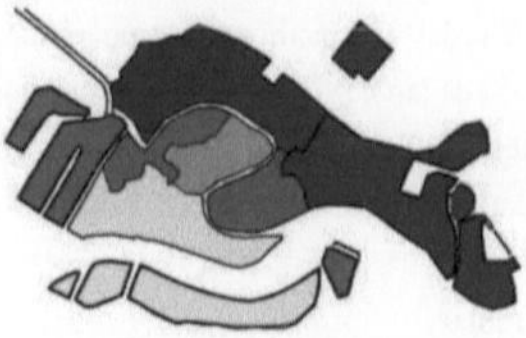

Sestieri of Venice:
Cannaregio
Castello
Dorsoduro
San Marco
San Polo
Santa Croce

These districts consist of parishes – initially seventy in 1033, but reduced under Napoleon and now numbering just thirty-eight. These parishes predate the sestieri, which were created in about 1170.

Other islands of the Venetian Lagoon do not form part of any of the sestieri, having historically enjoyed a considerable degree of autonomy.

Each sestiere has its own house numbering system. Each house has a unique number in the district, from one to several thousand, generally numbered from one corner of the area to another, but not usually in a readily understandable manner.

Aerial view of Venice

At the front of the Gondolas that work in the city there is a large piece of metal intended as a likeness of the Doge's hat. On this sit six notches pointing forwards and one pointing backwards. Each of these represent one of the Sestieri (the one that points backward represents the Giudecca).

Climate

According to the Köppen climate classification, Venice has a Humid subtropical climate (*Cfa*), with cool winters and very warm summers. The 24-hour average in January is 2.5 °C (**unknown operator: u'strong'** °F), and for July this figure is 22.7 °C (**unknown operator: u'strong'** °F). Precipitation is spread relatively evenly throughout the year, and averages 801 millimetres (**unknown operator: u'strong'** in).

Economy

Venice's economy has changed throughout history. In the Middle-Ages and the Renaissance, Venice was a major centre for commerce and trade, as it controlled a vast sea-empire, and became an extremely wealthy European city, a leader in political and economic affairs and a centre for trade and commerce.[38] Since the 11th century until the 15th century pilgrimages to the Holy Land were offered in Venice. Other ports such as Genoa, Pisa, Marseille, Ancona and Dubrovnik were hardly able to make any competition to the well organized transportation of pilgrims from Venice.[39] [40] This all changed by the 17th century, when Venice's trade empire was taken over by other countries such as Portugal, and its naval importance was reduced. In the 18th century, then, it became a major agricultural and industrial exporter. The 18th century's biggest industrial complex was the Venice Arsenal, and the Italian Army still uses it today (even though some space has been used for major theatrical and cultural productions, and beautiful spaces for art).[41] Today, Venice's economy is mainly based on tourism, shipbuilding (mainly done in the neighbouring cities of Mestre and Porto Marghera), services, trade and industrial exports.[38] Murano glass production in Murano and lace production in Burano are also highly important to the economy.[38]

Tourism

Venice is one of the most important tourist destinations in the world, due to the city being one of the world's greatest and most beautiful cities of art.[42] The city has an average of 50,000 tourists a day (2007 estimate).[43] In 2006, it was the world's 28th most internationally visited city, with 2.927 million international arrivals that year.[44]

Piazza San Marco. Doge's Palace

The Ponte dei Sospiri, the "Bridge of Sighs".

A gondola and a gondolier in the Grand Canal.

Tourism has been a major sector of Venetian industry since the 18th century, when it was a major center for the Grand Tour, due to its beautiful cityscape, uniqueness, and rich musical and artistic cultural heritage. In the 19th century, it became a fashionable centre for the rich and famous, often staying or dining at luxury establishments such as the Danieli Hotel and the Caffè Florian. It continued being a fashionable city in vogue right into the early 20th century.[42] In the 1980s, the Carnival of Venice was revived and the city has become a major centre of international conferences and festivals, such as the prestigious Venice Biennale and the Venice Film Festival, which attract visitors from all over the world for their theatrical, cultural, cinematic, artistic, and musical productions[42]

Today, there are numerous attractions in Venice, such as St Mark's Basilica, the Grand Canal, and the Piazza San Marco. The Lido di Venezia is also a popular international luxury destination, attracting thousands of actors, critics, celebrities, and mainly people in the cinematic industry. The city also relies heavily on the cruise business.[42]

However, Venice's popularity as a major worldwide tourist destination has caused several problems, including the fact that the city can be very overcrowded at some points of the year. It is regarded by some as a tourist trap, and by others as a 'living museum'.[42] Unlike most other places in Western Europe, and the world, Venice has become widely known for its element of elegant decay. The competition for foreigners to buy homes in Venice has made prices rise so highly that numerous inhabitants are forced to move to more affordable areas of Veneto and Italy, the most notable being Mestre.

Transport

Venice is built on an archipelago of 117 islands formed by 177 canals in a shallow lagoon, connected by 409 bridges.[45] In the old centre, the canals serve the function of roads, and almost every form of transport is on water or on foot. In the 19th century a causeway to the mainland brought the Venezia Santa Lucia railway station to Venice, and the Ponte della Libertà road causeway and parking facilities were built during the twentieth century. Beyond the road/rail land entrances at the northern edge of the city, transportation within the city remains (as it was in centuries past) entirely on water or on foot. Venice is Europe's largest urban car-free area. Venice is unique in Europe, in having remained a sizable functioning city in the twenty-first century entirely without motorcars or trucks.

Aerial view of Venice including the *Ponte della Libertà* bridge to the mainland

Waterways

The classical Venetian boat is the gondola, although it is now mostly used for tourists, or for weddings, funerals, or other ceremonies. Many gondolas are lushly appointed with crushed velvet seats and Persian rugs. Less well-known is the smaller sandolo. The main transportation means are motorised waterbuses (*vaporetti*), which ply regular routes along the major canals and between the city's islands, and private boats. The only gondolas still in common use by Venetians are the *traghetti*, foot passenger ferries crossing the Grand Canal at certain points without bridges.

Public transport

Azienda Consorzio Trasporti Veneziano (ACTV) is the name of the public transport system in Venice. It combines land transportation, with buses, and canal travel, with water buses (vaporetti). In total, there are 25 routes that connect the city.

The Venice People Mover (managed by ASM) is a cable operated public transit system connecting Tronchetto island with Piazzale Roma. Water taxis are also active.

Airports

Venice is served by the Marco Polo International Airport, or *Aeroporto di Venezia Marco Polo*, named in honor of its famous citizen. The airport is on the mainland and was rebuilt away from the coast. From the Venice airport, it's possible to reach by public transport:

- Venice Piazzale Roma by ATVO (provincial company) buses[46] and by ACTV (city company) buses (route 5 *aerobus*);[47]
- Venice, Lido and Murano by Alilaguna (private company) boats;
- Mestre, the mainland and Venice Mestre railways station (convenient for connections to Milan, Padova, Trieste, Verona and the rest of Italy) by ACTV lines (route 15 and 45)[47] and by ATVO lines;
- regional destination (Treviso, Padua, beach, ...) by ATVO buses and by Busitalia Sita Nord[48] buses (national company).

Morning Impression along a Canal in Venice, Veneto, Italy. by Rafail Sergeevich Levitsky.(1896) The Di Rocco Wieler Private Collection, Toronto, Canada

Some airlines market Treviso Airport in Treviso, 30 km from Venice, as a Venice gateway. Some simply advertise flights to "Venice", while naming the actual airport only in small print.[49] To reach Venice from Treviso airport people can catch a public bus from the company ATVO.

Venezia Lido,[50] a public airport suitable for smaller aircraft, is found on the NE end of Lido di Venezia. It has a 1000-metre grass runway.

Trains

Venice is serviced by regional and national trains, which can connect the city to Rome in 3.5 hours and to Milan in 2.5 hours. Treviso is thirty-five minutes away.[51] Florence and Padua are two of the stops between Rome and Venice. The St. Lucia station is a few steps away from a vaporetti stop.

The station is the terminus and starting point of the Venice Simplon Orient Express from or to London Victoria and Paris.

Car

The maritime portion of Venice has no roads as such, being composed almost entirely of narrow footpaths, and laid out across islands connected by staired stone footbridges, making transportation impossible by almost anything with wheels. Cars can reach the car/bus terminal via the Ponte della Libertà bridge. It comes in from the West from Mestre. There are two parking lots that serve the city: Tronchetto and Piazzale Roma. A ferry to Lido leaves from the parking lot in Tronchetto, and it is served by vaporetti and buses of the public transportation.

Administrative subdivision

The whole *comune* (English: municipality) *di Venezia* is divided into 6 *municipalità* (English: boroughs):

1) Venezia (*historic city*)-Murano-Burano (*Venezia insulare*): 69.136

2) Lido-Pellestrina (*Venezia litorale*): 21.664

Mainland (terraferma):

3) Favaro Veneto: 23.615

4) Mestre-Carpenedo (*Mestre centro*): 88.952

5) Chirignago-Zelarino: 38.179

6) Marghera: 28.466

A Royal Decree, in 1926, annexed mainland to the comune of Venezia.

Education

Venice is a major international centre for higher education. The city hosts Ca' Foscari University of Venice founded in 1868, Iuav University of Venice founded in 1926 and Venice International University an international research center founded in 1995 located on the island of San Servolo.

Demographics

In 2009, there were 270,098 people residing in Venice's comune (the population estimate of 272,000 inhabitants includes the population of the whole Comune of Venezia; around 60,000 in the historic city of Venice (Centro storico); 176,000 in Terraferma (the Mainland); and 31,000 live on other islands in the lagoon), of whom 47.4% were male and 52.6% were female. Minors (children ages 18 and younger) totalled 14.36 percent of the population compared to pensioners who number 25.7 percent. This compares with the Italian average of 18.06 percent (minors) and 19.94 percent (pensioners). The average age of Venice residents is 46 compared to the Italian average of 42. In the five years between 2002 and 2007, the population of Venice declined by 0.2 percent, while Italy as a whole grew by 3.85 percent.[52] But the population in the historic old city declines at a significantly faster rate: from about 120,000 in 1980 to about 60,000 in 2009.[53]

As of 2009, 91% of the population was Italian. The largest immigrant group comes from other European nations (Romanians, the largest group: 3%, South Asia: 1.3%, and East Asia: 0.9%). Venice is predominantly Roman Catholic, but because of the long standing relationship with Constantinople there is also a perceptible Orthodox presence, and due to immigration it now has some Muslim, Hindu and Buddhist inhabitants.

There is also a historic Jewish Community in Venice. The Venetian Ghetto was the area in which Jews were compelled to live under the Venetian Republic. It is from its name, in the Venetian language, that the word "ghetto", used in many languages, is derived. William Shakespeare's play *The Merchant of Venice*, probably written in the late 16th century, features Shylock, a Venetian Jew and his family. Venice also has an eruv,[54] built for and still used by the Jewish community.

Municipal administration

Composition of the City Council	
Party	**Members**
PD	17
PDL	10
Left and Greens	5
LN	4

- Name of the Mayor: Giorgio Orsoni
- Date of election: 30 March 2010
- Party: Democratic Party

Venice's City Council is composed by 45 members.

Of six boroughs into which Venice is divided, five are governed by centre-left coalition and one by centre-right coalition.

The current mayor of Venice, Giorgio Orsoni, has been elected in March 2010. He has been also *Assessore* since 2000 to 2005 and councilior of the Venice Biennale since 2000 to 2003.

Culture

Cinema and Venice in popular culture and media

Typical masks worn during the Carnival of Venice.

Venice has been the setting or chosen location of numerous films, novels, poems and other cultural references. The city was a particularly popular setting for several novels, essays, and other works of fictional or non-fictional literature. Examples of these include Shakespeare's *Merchant of Venice* and *Othello*, Ben Jonson's *Volpone*, Voltaire's *Candide*, Casanova's autobiographical *History of My Life*, Anne Rice's *Cry to Heaven*, and Philippe Sollers' *Watteau in Venice*. Thomas Mann's 1912 novella, *Death in Venice,* has served as the basis for an opera (Benjamin Britten's *Death in Venice*), a film (Visconti's *Death in Venice*) and a cocktail (*Death in Venice* [55]). The city has also been a setting for numerous other films such as the James Bond series *From Russia with Love*, *Moonraker*, *The Tourist*, *Casino Royale*, Katharine Hepburn in *Summertime*, *Fellini's Casanova*, *Don't Look Now*, *Indiana Jones and the Last*

Crusade, *A Little Romance*, *The Italian Job*, *Lara Croft: Tomb Raider*, and *The Talented Mr. Ripley*. The city has also been the setting for music videos such as Siouxsie and the Banshees' "Dear Prudence" and Madonna's *Like a Virgin (song)*, as well as in the video game *Assassin's Creed II*.

Architecture

Venice has a rich and diverse architectural style, the most famous of which is the Gothic style. Venetian Gothic architecture is a term given to a Venetian building style combining use of the Gothic lancet arch with Byzantine and Arab influences. The style originated in 14th-century Venice, where the confluence of Byzantine style from Constantinople met Arab influence from Moorish Spain. Chief examples of the style are the Doge's Palace and the Ca' d'Oro in the city. The city also has several Renaissance and Baroque buildings, including the Ca' Pesaro and the Ca' Rezzonico.

The Baroque Ca' Rezzonico

Music and the performing arts

The city of Venice in Italy has played an important role in the development of the music of Italy. The Venetian state – i.e., the medieval Maritime Republic of Venice – was often popularly called the "Republic of Music", and an anonymous Frenchman of the 17th century is said to have remarked that "In every home, someone is playing a musical instrument or singing. There is music everywhere."[56]

During the 16th century, Venice became one of the most important musical centers of Europe, marked by a characteristic style of composition (the Venetian school) and the development of the Venetian polychoral style under composers such as Adrian Willaert, who worked at St Mark's Basilica. Venice was the early center of music printing; Ottaviano Petrucci began publishing music almost as soon as this technology was available, and his publishing enterprise helped to attract composers from all over Europe, especially from France and Flanders. By the end of the century, Venice was famous for the splendor of its music, as exemplified in the "colossal style" of Andrea and Giovanni Gabrieli, which used multiple choruses and instrumental groups. Venice was also the home of many famous composers during the baroque period, such as Antonio Vivaldi, Ippolito Ciera, Giovanni Picchi, and Girolamo Dalla Casa, to name but a few.

La Fenice operahouse in the city

Interior design

It can be argued that Venice produced the most unique and refined Rococo designs. At the time, Venice was in a state of trouble. It had lost most of its maritime power, was lagging behind its rivals in political importance, and society had become decadent, with nobles wasting their money in gambling and partying. But Venice remained Italy's fashion capital, and was a serious contender to Paris in terms of wealth, architecture, luxury, taste, sophistication, trade, decoration, style, and design.[57] Venetian Rococo was well known for being rich and luxurious, with usually very extravagant designs. Unique Venetian furniture, such as the *divani da portego*, or long Rococo couches and *pozzetti*, objects meant to be placed against the wall. Venetian bedrooms were usually sumptuous and grand, with rich damask, velvet, and silk drapery and curtains, a beautifully carved Rococo beds with statues of putti, flowers and angels.[57] Venice was especially famous for its beautiful girandole mirrors, which remained among, if not, the finest in Europe. Chandeliers were usually very colourful, using Murano glass to make them look more vibrant and stand out from others, and precious stones and materials from abroad were used, since Venice still held a vast trade empire. Lacquer was very common, and many items of furniture were covered with it, the most famous being *lacca povera* (poor lacuqer), in which allegories and images of social life were painted. Lacquerwork and Chinoiserie were particularly common in bureau cabinets.[58]

Fashion and shopping

Luxury shops and boutiques along the Rialto Bridge.

In the 14th century, many young Venetian men began wearing tight-fitting multicoloured hose, the designs on which indicated the Compagnie della Calza ("Trouser Club") to which they belonged. The Venetian Senate passed sumptuary laws, but these merely resulted in changes in fashion in order to circumvent the law. Dull garments were worn over colourful ones, which then were cut to show the hidden colours resulting in the wide spread of men's "slashed" fashions in the 15th century.

Today, Venice is also a major fashion and shopping centre in Italy, not as important as Milan, Florence, or Rome, but par to Turin, Vicenza, Naples, and Genoa. Roberta di Camerino is the only major Italian fashion brand to be based out of Venice.[59] Founded in 1945, it is renowned for its innovative handbags featuring hardware by Venetian artisans and often covered in locally woven velvet, and has been credited with creating the concept of the easily recognisable status bag.[59] Many of the fashion boutiques and jewelry shops in the city are located in the Rialto Bridge and the Piazza San Marco. At the current time, there are Louis Vuitton and Ermenegildo Zegna flagship stores operating in the city.

Cuisine

Venetian cuisine is characterized by seafood, but also includes garden products from the islands of the lagoon, rice from the mainland, game, and polenta. Venice combines local traditions with influences that are distant from millennial business contacts. These include *sarde in saor*, sardines marinated in order to preserve them for long voyages; *risi e bisi*, rice, peas and ham; *fegato alla veneziana*, Venetian-style liver; risotto with cuttlefish, blackened from the ink; *cicchetti*, refined and delicious tidbits (akin to *tapas*); *antipasti*, appetizers; and *prosecco*, an effervescent, mildly sweet wine.

Hot chocolate was a fashionable drink in Venice during the 1770s and 1780s.

In addition, Venice is famous for *bisàto* (marinated eel), for the golden, oval-shaped cookies called baicoli, and for different types of sweets such as: *pan del pescatore* (bread of the fisherman); cookies with almonds and pistachio nuts; cookies with fried Venetian cream or the *bussolai* (butter biscuits and shortbread made in the shape of an "S" or ring) from the island of Burano; the *crostoli* also known as the chatter, lies, or *galani*; the *fregolotta* (a crumbly cake with almonds); milk pudding called *rosada*; and cookies of yellow semolina called *zaléti*.

Language

Venetian or the regional form Venetan is a Romance language spoken as a native language by over two million people,[60] mostly in Venice, but also the Veneto region of Italy, where of five million inhabitants almost all can understand it. It is sometimes spoken and often well understood outside Veneto, in Trentino, Friuli, Venezia Giulia, Istria, and some towns of Dalmatia, an area of six to seven million people. The language enjoyed substantial prestige in the days of the Venetian Republic, when it attained the status of a lingua franca in the Mediterranean.

Literature

Venice has long been a source of inspiration for authors, poets and playwrights as well as being at the forefront of the technical developing of printing and publishing.

Portrait of Giacomo Casanova

Two of the most famous Venetian writers were Marco Polo in the Middle Ages and later Giacomo Casanova. Polo (1254–1324) was a merchant who voyaged to the Orient. His series of books, co-written by Rustichello da Pisa, titled *Il Milione* provided important knowledge of the lands east of Europe, from the Middle East, to China, Japan and Russia. Giacomo Casanova (1725–1798) was a prolific writer and famous adventurer best remembered for his autobiography, *Histoire De Ma Vie* (Story of My Life), which links his colourful lifestyle to the city of Venice.

Venetian playwrights followed the old Italian theatre tradition of *Commedia dell'Arte*. Ruzante (1502–1542) and Carlo Goldoni (1707–1793) used the Venetian dialect extensively in their comedies.

book printed by Aldus Manutius

Venice has also inspired writers from abroad. Shakespeare set Othello and The Merchant of Venice in the city. Thomas Mann authored the novel *Death in Venice*, published in 1912. Venice inspired the poetry of Ezra Pound, who wrote his first literary work in the city. Pound died in 1972 and his remains are buried in Venice's cemetery island of San Michele. The French writer Philippe Sollers spent most of his life in Venice and published *A Dictionary For Lovers Of Venice* in 2004. Ugo Foscolo (1778–1827) born in Zante, an island that at the time belonged to the Republic of Venice, was also a famous poet and revolutionary who wanted to see a free republic established in Venice following the fall to Napoleon. The city features prominently in Henry James' The Wings of the Dove and is also visited in Evelyn Waugh's Brideshead Revisited and Marcel Proust's In Search of Lost Time.

Venice is also linked to the technical aspects of writing. The city was the location for one of Italy's earliest printing presses, established by Aldus Manutius (1449–1515). From this beginning Venice developed as an important typographic center and even as late as the 18th century was responsible for printing half of Italy's published books.

Art and printing

An 18th century view of Venice by Venetian artist Canaletto.

Venice, especially during the Middle-Ages, Renaissance and Baroque, was a major centre of art and developed a unique style known as the Venetian School. In the Middle-Ages and the Renaissance, Venice, along with Florence and Rome, became one of the most important centres of art in Europe, and numerous wealthy Venetians became patrons of the arts. Venice at the time was a rich and prosperous Maritime Republic, which controlled a vast sea and trade empire.[61]

By the end of the 15th century, Venice had become the European capital of printing, being one of the first cities in Italy (after Subiaco and Rome) to have a printing press after those established in Germany, having 417 printers by 1500. The most important printing office was the Aldine Press of Aldus Manutius, which in 1499 printed the Hypnerotomachia Poliphili, considered the most beautiful book of Renaissance, and established modern punctuation, the page format and italic type, and the first printed work of Aristotle.

In the sixteenth century Venetian painting was developed through influences from the Paduan School and Antonello da Messina, who introduced the oil painting technique of the van Eyck brothers. It is signified by a warm colour scale and a picturesque use of colour. Early masters where the Bellini and Vivarini families, followed by Giorgione and Titian, then Tintoretto and Veronese. In the early 16th century, also, there was rivalry between whether Venetian painting should use *disegno* or *colorito*.[62]

Canvases (the common painting surface) originated in Venice during the early renaissance. These early canvases were generally rough.

In the eighteenth century Venetian painting had a renaissance because of Tiepolo's decorative painting and Canaletto's and Guardi's panoramic views.

Glass

A Venetian glass goblet

Venice is famous for its ornate glass-work, known as Venetian glass. It is world-renowned for being colourful, elaborate, and skilfully made.

Many of the important characteristics of these objects had been developed by the thirteenth century. Toward the end of that century, the center of the Venetian glass industry moved to Murano.

Byzantine craftsmen played an important role in the development of Venetian glass, an art form for which the city is well-known. When Constantinople was sacked by the Fourth Crusade in 1204, some fleeing artisans came to Venice. This happened again when the Ottomans took Constantinople in 1453, supplying Venice with still more glassworkers. By the sixteenth century, Venetian artisans had gained even greater control over the color and transparency of their glass, and had mastered a variety of decorative techniques.

An ornate Murano glass chandelier.

Despite efforts to keep Venetian glassmaking techniques within Venice, they became known elsewhere, and Venetian-style glassware was produced in other Italian cities and other countries of Europe.

Some of the most important brands of glass in the world today are still produced in the historical glass factories on Murano. They are: Venini, Barovier & Toso, Pauly, Millevetri, Seguso.[63] Barovier & Toso is considered one of the 100 oldest companies in the world, formed in 1295.

One of the most renowned types of Venetian glasses are made in Murano, known as Murano glass, which has been a famous product of the Venetian island of Murano for centuries. Located off the shore of Venice, Italy, Murano was a commercial port as far back as the 7th century. By the 10th century it had become a well-known city of trade. Today Murano remains a destination for tourists and art and jewellery lovers alike.

Festivals

The Carnival of Venice is held annually in the city, starting around two weeks before Ash Wednesday and ends on Shrove Tuesday. The carnival is closely associated with Venetian masks.

The Venice Biennale is one of the most important events in the arts calendar. During 1893 headed by the mayor of Venice, Riccardo Selvatico, the Venetian City Council passed a resolution on 19 April to set up an Esposizione biennale artistica nazionale (biennial exhibition of Italian art), to be inaugurated on 22 April 1895.[64] Following the outbreak of hostilities during the Second World War, the activities of the Biennale were interrupted in September 1942, but resumed in 1948.[65]

The Festa del Redentore is held in mid July. It began as a feast to give thanks for the end of the terrible plague of 1576. A bridge of barges is built connecting Giudecca to the rest of Venice, and fireworks play an important role.

The Venice Film Festival (Italian *Mostra Internazionale d'Arte Cinematografica di Venezia*) is the oldest film festival in the world. Founded by Count Giuseppe Volpi di Misurata in 1932 as the "Esposizione Internazionale d'Arte Cinematografica", the festival has since taken place every year in late August or early September on the island of the Lido, Venice, Italy. Screenings take place in the historic Palazzo del Cinema on the Lungomare Marconi. It is one of the world's most prestigious film festivals and is part of the Venice Biennale.

Foreign words of Venetian origin

Words with a Venetian etymology include arsenal, ciao, ghetto, gondola, imbroglio, lagoon, lazaret, lido, Montenegro, quarantine, regatta. The name of Venezuela is a Spanish diminutive of Venice. Many other places around the world are named after Venice, e.g. Venice Beach.

Notable people

For people from Venice, see People from Venice. Others closely associated with the city include:

- Enrico Dandolo (*c.* 1107, 1205), Doge of Venice from 1192 to his death. He played a direct role in the Sack of Constantinople during the Fourth Crusade.
- Marco Polo (15 September 1254 – 8 January 1324), trader and explorer, one of the first Westerners to travel the Silk Road to China. While a prisoner in Genoa, he dictated in the tale of his travels known as *Il Milione* (*The Travels of Marco Polo*).
- Giovanni Bellini (c. 1430–1516), a Renaissance painter, probably the best known of the Bellini family of painters.
- Aldus Manutius (1449–1515), one of the most important printers in history.

- Pietro Bembo (20 May 1470 – 18 January 1547), cardinal and scholar.
- Lorenzo Lotto (c.1480 – Loreto, 1556), painter, draughtsman, and illustrator, traditionally placed in the Venetian school.
- Sebastian Cabot (c. 1484–1557, or soon after), explorer.
- Pellegrino Ernetti, Catholic priest and exorcist
- Titian (*c.* 1488–90 – 27 August 1576), leader of the 16th century Venetian school of the Italian Renaissance (he was born in Pieve di Cadore).
- Sebastiano Venier, (c. 1496 – 3 March 1578), Doge of Venice from 11 June 1577 to 1578.
- Andrea Gabrieli (c.1510–1586), Italian composer and organist at St Mark's Basilica
- Tintoretto (1518 – 31 May 1594), probably the last great painter of Italian Renaissance.
- Veronica Franco (1546–1591), poet and courtesan during the Renaissance
- Giovanni Gabrieli (between 1554 and 1557–1612), composer and organist at St Mark's Basilica
- Claudio Monteverdi (1567–1643), composer and director of music at San Marco
- Leon Modena (1571–1648) preacher, author, poet, active in the Venetian ghetto and beyond
- Marco Antonio Bragadin (d.1571), general, flayed alive by the Turks after a fierce resistance during the siege of Famagusta
- Baldassare Longhena (1598 – 18 February 1682), one of the greatest exponents of Baroque architecture.
- Francesco Cavalli (14 February 1602 – 14 January 1676), a baroque composer
- Pietro Cesare Alberti (1608–1655), considered the first Italian-American, arriving in New Amsterdam in 1635.
- Tomaso Albinoni (8 June 1671 – 17 January 1751), a baroque composer
- Rosalba Carriera (7 October 1675 – 15 April 1757), known for her pastel works.
- Antonio Vivaldi (4 March 1678, 28 July (or 27), 1741, Vienna), famous composer and violinist of the Baroque Era
- Pietro Guarneri (14 April 1695 – 7 April 1762) left Cremona in 1718, settled in Venice. "Peter of Venice" from the family of great luthiers.
- Giovanni Battista Tiepolo (5 March 1696 – 27 March 1770), the last "Grand Manner" fresco painter from the Venetian Republic.
- Canaletto (28 October 1697 – 19 April 1768), famous for his landscapes or *vedute* of Venice, but not only.
- Carlo Goldoni (25 February 1707 – 6 February 1793). Along with Pirandello, Goldoni is probably the most famous name in Italian theatre, in his country and abroad.
- Carlo Gozzi (13 December 1720 – 4 April 1806), an excellent dramatist of 18th century.
- Giacomo Casanova (1725–1798), in Dux, Bohemia, (now Duchcov, Czech Republic), a famous Venetian adventurer, writer and womanizer.
- Lorenzo Da Ponte (1749–1838), opera librettist and poet. He wrote the librettos for 28 operas by 11 composers, including Wolfgang Amadeus Mozart.
- Virgilio Ranzato (7 May 1883 – 20 April 1937), Composer.
- Carlo Scarpa (2 June 1906 – 1978, Sendai, Japan), an architect with a profound understanding of materials.
- Emilio Vedova (9 August 1919 – 25 October 2006), one of the most important modern painters of Italy
- Elena Lucrezia Cornaro Piscopia (5 June 1646 – 26 July 1684), the first woman in the world to receive a doctorate degree.
- Bruno Maderna (21 April 1920 – 13 November 1973), an Italian-German orchestra director and 20th century music composer.
- Luigi Nono (29 January 1924 – 8 May 1990), a leading composer of instrumental and electronic music.
- Ludovico de Luigi (November 1933), Venetian Surrealistic artist.
- Giuseppe Sinopoli (2 November 1946 – 20 April 2001), conductor and composer.
- Romano Scarpa (27 September 1927, Venice – 23 April 2005, Málaga), was one of the most famous Italian creators of Disney comics.

International relations

The City of Venice and the Central Association of Cities and Communities of Greece (KEDKE) established, in January 2000, in pursuance of the EC Regulations n. 2137/85, the European Economic Interest Grouping (E.E.I.G.) Marco Polo System to promote and realise European projects within transnational cultural and tourist field, particularly referred to the artistic and architectural heritage preservation and safeguard.

Twin towns and sister cities

Venice is twinned with:

• Esfahan, Iran • Islamabad, Pakistan, since 1960 • Palembang, Indonesia. • Suzhou, China, since 1980 • Bangkok,Thailand	• Tallinn, Estonia • Pereira, Colombia • Istanbul, Turkey, since 1993 • Sarajevo, Bosnia and Herzegovina, since 1994 • Yerevan, Armenia	• Aveiro, Portugal, since 1998 • Nuremberg, Germany, since 1999 • Qingdao, China, since 2001 • Saint Petersburg, Russia, since 2002	• Thessaloniki, Greece, since 2003 • Fort Lauderdale, United States, since 2007 • Wolverhampton, United Kingdom • Most, Czech Republic

Cooperation agreements

Venice has cooperation agreements with the Greek city of Thessaloniki, the German city of Nuremberg, signed on 25 September 1999, and the Turkish city of Istanbul, signed on 4 March 1993, within the framework of the 1991 Istanbul Declaration. It is also a Science and Technology Partnership City with Qingdao, China.

Etymology

The name is connected with the people known as the Veneti, perhaps the same as the *Eneti* (Ενετοί). The meaning of the word is uncertain. Connections with the Latin verb 'venire' (to come) or venia are fanciful. A connection with the Latin word venetus, meaning 'sea-blue', is possible.

See also

- Outline of Italy
- List of architecture monuments of Venice
- List of painters and architects of Venice
- Venetian Ghetto
- Jewish Community of Venice
- *Su e zo per i ponti*
- Veneti and Venetic language (the ancient spoken language of the region)
- S.S.C. Venezia
- Venetian Blinds
- Venetian language (the modern spoken vernacular of the region)
- Venezia Mestre Rugby FC – rugby team

Several cities have been compared to Venice: The city of Nantes, France, has been called *The Venice of the West*, many places have been named *Venice of the East*, while equally as many have been called *Venice of the North.*

References

Notes

[1] http://http://www.comune.venezia.it/

[2] http://whc.unesco.org/en/list/394

[3] http://whc.unesco.org/en/list

[4] http://whc.unesco.org/en/list/?search=&search_by_country=&type=&media=&region=&order=region

[5] UNESCO: Venice and its Lagoon (http://whc.unesco.org/en/list/394), accessed:17th April, 2012

[6] Mara Rumiz, Venice Demographics Official Mock funeral for Venice's 'death' (http://news.bbc.co.uk/1/hi/world/europe/8360253.stm)

[7] "Online Etymology Dictionary" (http://www.etymonline.com/index.php). . Retrieved 11 June 2010.

[8] Richard Stephen Charnock (1859). *Local etymology: a derivative dictionary of geographical names*. Houlston and Wright. p. 288.

[9] Barzini, Luigi (30 May 1982). "The Most Beautiful and Wonderful City In The World – The" (http://query.nytimes.com/gst/fullpage.html?res=950CE5DD1038F933A05756C0A964948260&sec=&spon=&pagewanted=2). *New York Times*. . Retrieved 28 March 2009.

[10] Bleach, Stephen; Schofield, Brian; Crump, Vincent (17 June 2007). "Europes most romantic city breaks" (http://www.timesonline.co.uk/tol/travel/holiday_type/breaks/article1936951.ece). *The Times* (London). . Retrieved 27 May 2010.

[11] "Venetian Music of the Renaissance" (http://www.vanderbilt.edu/Blair/Courses/MUSL242/f98/venice.htm). Vanderbilt.edu. 11 October 1998. . Retrieved 22 April 2010.

[12] "Imperciocchè nascendi i principati", begins Apostolo Zeno, *Compendio della storia Veneta di Apostolo Zeno continuata fino alla caduta della repubblica* 1847:9.

[13] Bosio, *Le origini di Venezia*

[14] Zeno, *Compendio* 1847:10.

[15] Trudy Ring; Robert M. Salkin; Sharon La Boda (1 January 1996). *International Dictionary of Historic Places: Southern Europe* (http://books.google.com/books?id=74JI2UlcU8AC&pg=PA745). Taylor & Francis. p. 745. ISBN 978-1-884964-02-2. . Retrieved 24 March 2011.

[16] Traditional date as given in William J. Langer, ed. *An Encyclopedia of World History*.

[17] According to John Julius Norwich, the traditional first doge of Venice, Paolo Lucio Anafesto, was actually Exarch Paul.

[18] John Julius Norwich, *A History of Venice* (Alfred A. Knopf: New York, 1982) p. 13.

[19] Langer.

[20] Richard Cowen, The importance of salt (http://www-geology.ucdavis.edu/~cowen/~gel115/salt.html)

[21] Herrin, Byzantium: The Surprising Life of a Medieval Empire, Penguin, Harmondsworth, ISBN 978-0-14-103102-6

[22] "History of Venice" (http://www.historyworld.net/wrldhis/PlainTextHistories.asp?ParagraphID=fqa). Historyworld.net. . Retrieved 28 March 2009.

[23] James Burke, *Connections* (Little, Brown and Co., 1978/1995, ISBN 978-0-316-11672-5, p.105

[24] William J. Bernstein (2009). " *A Splendid Exchange: How Trade Shaped the World* (http://books.google.com/books?id=ePiReZWp0NwC&pg=&dq=&hl=en#v=onepage&q=&f=false)". Grove Press. p.143. ISBN 0-8021-4416-0

[25] State of Texas, Texas Department of State Health Services. "History of Plague" (http://www.dshs.state.tx.us/preparedness/bt_public_history_plague.shtm). Dshs.state.tx.us. . Retrieved 28 March 2009.

[26] " *Medicine and society in early modern Europe* (http://books.google.com/books?id=fQxAkrbksTEC&pg=PA41&dq&hl=en#v=onepage&q=&f=false)". Mary Lindemann (1999). Cambridge University Press. p.41. ISBN 0-521-42354-6

[27] Italian site about *Schiavoni* (http://www.schiavoni.org/Storia.html)

[28] Patrick G. Skelly, Pocasset MA (6 May 2005). "US Army Air Force Operations Mediterranean Theater" (http://www.milhist.net/usaaf/mto.html). Milhist.net. . Retrieved 27 July 2010.

[29] Patrick G. Skelly, Pocasset MA (21 July 1945). "New Zealand troops relieve Venice" (http://www.milhist.net/history/onemoreriver.html). Milhist.net. . Retrieved 28 March 2009.

[30] Kendall, Paul (25 August 2010). "Mythology and Folklore of the Alder" (http://www.treesforlife.org.uk/forest/mythfolk/alder.html). Trees for life. . Retrieved 6 August 2011.

[31] "Alder - Alnus glutinosa" (http://www.cvni.org/treenursery/trees/alder). Conservation Volunteers Northern Ireland. . Retrieved 6 August 2011.

[32] Interview with Leonid Grigoriev (http://en.civilg8.ru/priority/energy/2054.php)

[33] "Venice turpentine" (http://www.darwinprice.com/venturp.htm). darwinprice.com. . Retrieved 6 August 2011.

[34] *Technology: Venetians put barrage to the test against the Adriatic* (http://web.archive.org/web/20071011072114/http://media.newscientist.com/article/mg12216602.900-technology-venetians-put-barrage-to-the-test-against-theadriatic-.html). New Scientist magazine. 15 April 1989. Archived from the original (http://media.newscientist.com/article/mg12216602.900-technology-venetians-put-barrage-to-the-test-against-theadriatic-.html) on 11 October 2007. . Retrieved 10 October 2007.

[35] "Venice's 1,500-year battle with the waves" (http://news.bbc.co.uk/2/hi/europe/3069305.stm). BBC News. 17 July 2003. . Retrieved 10 October 2007.

[36] "'Moses project' to secure future of Venice" (http://www.telegraph.co.uk/news/worldnews/europe/italy/3629387/Moses-project-to-secure-future-of-Venice.html). Telegraph News. 11 January 2012. . Retrieved 11 January 2012.

[37] "Tabella Climo" (http://www.meteoam.it/modules.php?name=viewClinoFile&station=105&name_station=Venezia Tessera). MeteoAM. June 2011. . Retrieved 29 May 2009.
[38] "The economy of Venice, Italy" (http://www.aboutvenice.org/economy-of-venice.html). Aboutvenice.org. . Retrieved 22 April 2010.
[39] Pilgerreisen von Venedig nach Jerusalem im späten Mittelalter- Die Verträge mit dem Schiffspatron, Seite 2, Fabian H. Flöper, GRIN Verlag, 2011. ISBN 9783656047834
[40] Venice, page 71, Beryl D. De Sélincourt, May (Sturge) Gretton, Chatto & Windus, London 1907., reprinted BiblioBazaar 2010, ISBN 9781177404488
[41] "Venice (Italy) :: Economy – Britannica Online Encyclopedia" (http://www.britannica.com/EBchecked/topic/625298/Venice/24381/Economy). Britannica.com. . Retrieved 22 April 2010.
[42] "Venice (Italy) :: Economy – Britannica Online Encyclopedia" (http://www.britannica.com/EBchecked/topic/625298/Venice/24381/Economy#). Britannica.com. . Retrieved 22 April 2010.
[43] Owen, Richard (5 April 2007). "Venice in peril as the tourists flood in and locals get out" (http://www.timesonline.co.uk/tol/travel/destinations/italy/article1615074.ece). *The Times* (London). . Retrieved 27 May 2010.
[44] "Top 150 City Destinations London Leads the Way > Euromonitor archive" (http://www.euromonitor.com/Top_150_City_Destinations_London_Leads_the_Way). Euromonitor.com. 11 October 2007. . Retrieved 22 April 2010.
[45] "Venice Study Abroad" (http://www.studyabroad.com/pages/sitecontent/venice.aspx). . Retrieved 6 October 2010.
[46] http://www.atvo.it/index.php?lang=it&area=23&menuid=35
[47] http://www.actv.it/muoversiinterraferma/lineeurbane
[48] http://www.fsbusitalia.it/cms-instance/documenti/fsbusitalia/Montegrotto-AeropMarcoPolo.pdf
[49] Home Page (http://wizzair.com/default.asp?slid=clear&language=EN)", *Wizz Air*
[50] http://www.enav.it/enavWebPortalStatic/AIP/AD/AD2/ADSPV1-1.pdf
[51] *Thomas Cook European Timetables*
[52] "Statistiche demografiche ISTAT" (http://demo.istat.it/bil2007/index.html). Demo.istat.it. . Retrieved 28 March 2009.
[53] Cathy Newman, "*Vanishing Venice*", National Geographic, August 2009
[54] "Venetian Ghetto – Eruv in Venice" (http://www.ghetto.it/ghetto/en/foto.asp?padre=3&figlio=1). . Retrieved 2 August 2010.
[55] http://www.thevenicelido.com/2011/08/death-in-venice-cocktail.html
[56] Touring Club p. 79
[57] Miller (2005) p.82
[58] Miller (2005) p.83
[59] Patner, Josh (26 February 2006). "From Bags to Riches" (http://www.nytimes.com/2006/02/26/style/tmagazine/t_w_p166_talk_patner_.html). *The New York Times*. . Retrieved 14 May 2010.
[60] Ethnologue.
[61] "The Renaissance in Venice – Art History Basics on the Venetian School – ca 1450–1600" (http://arthistory.about.com/cs/arthistory10one/a/ven_ren.htm). Arthistory.about.com. 29 October 2009. . Retrieved 22 April 2010.
[62] "Venetian art around 1500" (http://www.webexhibits.org/feast/context/venetianart.html). Webexhibits.org. . Retrieved 22 April 2010.
[63] Carl I. Gable, *Murano Magic: Complete Guide to Venetian Glass, its History and Artists* (Schiffer, 2004). ISBN 978-0-7643-1946-4.
[64] "The Venice Biennale: History of the Venice Biennale" (http://www.labiennale.org/en/biennale/history/origin/en/7823.html). Labiennale.org. . Retrieved 28 March 2009.
[65] "The Venice Biennale: History From the beginnings until the Second World War (1893–1945)" (http://www.labiennale.org/en/biennale/history/origin/en/7823.3.html). Labiennale.org. . Retrieved 28 March 2009.

Bibliography

Academic

- Bosio, Luciano. *Le origini di Venezia*. Novara: Istituto Geografico De Agostini.
- Brown, Horatio, *Venice*, chapter 8 of *Cambridge Modern History* vol. I *The Renaissance* (1902)
- Brown, Horatio, *Calendar of State Papers (Venetian): 1581–1591*, 1895; *1592–1603*, 1897; *1603–1607*, 1900; *1607–1610*, 1904; *1610–1613*, 1905
- Brown, Horatio, *Studies in the history of Venice* (London, 1907)
- Chambers, D.S. (1970). *The Imperial Age of Venice, 1380–1580.* London: Thames & Hudson. The best brief introduction in English, still completely reliable.
- Contarini, Gasparo (1599). *The Commonwealth and Gouernment of Venice.* Lewes Lewkenor, trsl. London: "Imprinted by I. Windet for E. Mattes." The most important contemporary account of Venice's governance during the time of its blossoming. Also available in various reprint editions.
- Da Canal, Martin, "Les estoires de Venise" (13th century chronicle), translated by Laura Morreale. Padua, Unipress 2009.

- Drechsler, Wolfgang (2002). "Venice Misappropriated." *Trames* 6(2), pp. 192–201. A scathing review of Martin & Romano 2000; also a good summary on the most recent economic and political thought on Venice.
- Garrett, Martin, "Venice: a Cultural History" (2006). Revised edition of "Venice: a Cultural and Literary Companion" (2001).
- Grubb, James S. (1986). "When Myths Lose Power: Four Decades of Venetian Historiography." *Journal of Modern History* 58, pp. 43–94. The classic "muckraking" essay on the myths of Venice.
- Lane, Frederic Chapin. *Venice: Maritime Republic* (1973) (ISBN 978-0-8018-1445-7) standard scholarly history; emphasis on economic, political and diplomatic history
- Laven, Mary, "Virgins of Venice: Enclosed Lives and Broken Vows in the Renaissance Convent (2002). The most important study of the life of Renaissance nuns, with much on aristocratic family networks and the life of women more generally.
- Martin, John Jeffries and Dennis Romano (eds). *Venice Reconsidered. The History and Civilization of an Italian City-State, 1297–1797.* (2002) Johns Hopkins University Press. The most recent collection on essays, many by prominent scholars, on Venice.
- Muir, Edward (1981). *Civic Ritual in Renaissance Venice.* Princeton UP. The classic of Venetian cultural studies, highly sophisticated.
- Oppenheimer, Gerald J. (2010). *Venetian* Palazzi *and* Case: *A Guide to the Literature.* University of Washington, Seattle. Retrieved from http://faculty.washington.edu/gerryo/venice.html 7 February 2010.
- Rösch, Gerhard (2000). *Venedig. Geschichte einer Seerepublik.* Stuttgart: Kohlhammer. In German, but the most recent top-level brief history of Venice.
- Miller, Judith (2005). *Furniture: world styles from classical to contemporary.* DK Publishing. ISBN 978-0-7566-1340-2.

Popular

- Ackroyd, Peter. *Venice: Pure City.* London, Chatto & Windus. 2009. ISBN 978-0-7011-8478-0
- Brown, Horatio, *Life on the Lagoons*, 1884; revised ed. 1894; further eds. 1900, 1904, 1909.
- Cole, Toby. *Venice: A Portable Reader,* Lawrence Hill, 1979. ISBN 978-0-88208-097-0 (hardcover); ISBN 978-0-88208-107-6 (softcover).
- Morris, Jan (1993), *Venice.* 3rd revised edition. Faber & Faber, ISBN 978-0-571-16897-2. A subjective and passionate written introduction to the city and some of its history. Not illustrated.
- Ruskin, John (1853). *The Stones of Venice.* Abridged edition Links, JG (Ed), Penguin Books, 2001. ISBN 978-0-14-139065-9. Seminal work on architecture and society
- di Robilant, Andrea (2004). *A Venetian Affair.* Harper Collins. ISBN 978-1-84115-542-5 Biography of Venetian nobleman and lover, from correspondence in the 1750s.
- Sethre, Janet. *The Souls of Venice* McFarland & Company, Inc., 2003. ISBN 978-0-7864-1573-1 (softcover). This book focuses on people who have been shaped by Venice and who have shaped the city in their turn. Illustrated (photographs by Manuela Fardin).

External links

- Venice travel guide from Wikitravel
- Venipedia (http://venipedia.org/index.php?title=Main_Page) wiki dedicated to Venice
- Official Site of the City of Venice (http://www.comune.venezia.it/flex/cm/pages/ServeBLOB.php/L/EN/IDPagina/1)
- Venice European Capital of Culture 2019 (http://www.nordest2019.eu/)
- Geschichte Venedigs (http://www.geschichte-venedigs.de/politik.html) (German)
- Archivio di Stato di Venezia (http://www.archiviodistatovenezia.it/) (Italian)
- Fondazione Musei Civici di Venezia (http://www.museiciviciveneziani.it/) (Italian/English)
- Fondazione Giorgio Cini (http://www.cini.it/) (Italian/English)

Switzerland

Motto: *Unus pro omnibus, omnes pro uno* (traditional)
(Latin: One for all, all for one)

Anthem: Swiss Psalm

Location of Switzerland(green)
in Europe(dark grey) — [Legend]

Capital	Bern[1] (de facto) 46°57′N 7°27′E
Largest city	Zurich
Official language(s)	German (63.7 %), French (20.4 %), Italian (6.5 %), Romansh[2] (0.5 %)
Demonym	Swiss
Government	Federal republic, with directorial system and direct democracy
- Federal Council	Doris Leuthard Eveline Widmer-Schlumpf (Pres. 12) Ueli Maurer (VP 12) Didier Burkhalter Simonetta Sommaruga Johann Schneider-Ammann Alain Berset
- Federal Chancellor	Corina Casanova
Legislature	Federal Assembly

-	Upper house	Council of States
-	Lower house	National Council
Independence		
-	Foundation date	1 August[3] 1291
-	*de facto*	22 September 1499
-	Recognised	24 October 1648
-	Restored	7 August 1815
-	Federal state	12 September 1848[4]
Area		
-	Total	41,285 km^2 (133rd) 15,940 sq mi
-	Water (%)	4.2
Population		
-	2011 estimate	7,952,600[5] (95th)
-	2000 census	7,452,075
-	Density	188/km^2 (65th) 477.4/sq mi
GDP (PPP)		2011 estimate
-	Total	$339.890 billion[6]
-	Per capita	$43,369[6]
GDP (nominal)		2011 estimate
-	Total	$636.059 billion[6]
-	Per capita	$81,160[6]
Gini (2000)		33.7 (medium)
HDI (2011)		▲ 0.903[7] (very high) (11th)
Currency		Swiss franc (CHF)
Time zone		CET (UTC+1)
-	Summer (DST)	CEST (UTC+2)
Drives on the		right (trains: left)
ISO 3166 code		CH
Internet TLD		.ch
Calling code		+41

Switzerland (German: *die Schweiz*[8] [ˈʃvaɪts]; French: *Suisse* [sɥis(ə)]; Italian: *Svizzera* Italian pronunciation: [ˈzviːtːsɛra]; Romansh: *Svizra* [ˈʒviːtsrɐ] or [ˈʒviːtsʁːɐ]), in its full name the **Swiss Confederation** (Latin: *Confoederatio Helvetica*, hence its abbreviation CH), is a federal republic consisting of 26 cantons, with Bern as the seat of the federal authorities. The country is situated in western Europe,[9] where it is bordered by Germany to the north, France to the west, Italy to the south, and Austria and Liechtenstein to the east.

Switzerland is a landlocked country geographically divided between the Alps, the Swiss Plateau and the Jura, spanning an area of 41285 km^2 (**unknown operator: u'strong'** sq mi). While the Alps occupy the greater part of the territory, the Swiss population of approximately 8 million people is concentrated mostly on the Plateau, where the largest cities are to be found. Among them are the two global cities and economic centres of Zurich and Geneva.

The Swiss Confederation has a long history of armed neutrality—it has not been in a state of war internationally since 1815—and did not join the United Nations until 2002. It pursues, however, an active foreign policy and is frequently involved in peace-building processes around the world.[10] Switzerland is also the birthplace of the Red Cross and home to a large number of international organizations, including the second largest UN office. On the European level, it is a founding member of the European Free Trade Association and is part of the Schengen Area – although it is notably not a member of the European Union, nor the European Economic Area.

Switzerland is one of the richest countries in the world by per capita gross domestic product, and has the highest wealth per adult (financial and non-financial assets) of any country in the world.[11] [12] Zurich and Geneva have respectively been ranked as the cities with the second and third highest quality of life in the world.[13] It has the world's nineteenth largest economy by nominal GDP and the thirty-sixth largest by purchasing power parity. It is the twentieth largest exporter and eighteenth largest importer of goods.

Switzerland comprises three main linguistic and cultural regions: German, French, and Italian, to which the Romansh-speaking valleys are added. The Swiss, therefore, though predominantly German-speaking, do not form a nation in the sense of a common ethnic or linguistic identity. The strong sense of belonging to the country is founded on the common historical background, shared values (federalism and direct democracy)[14] and Alpine symbolism.[15] The establishment of the Swiss Confederation is traditionally dated to 1 August 1291; Swiss National Day is celebrated on the anniversary.

Etymology

The English name *Switzerland* is a compound containing *Switzer*, an obsolete term for the Swiss, which was in use during the 16th to 19th centuries.[16] The English adjective *Swiss* is a loan from French *Suisse*, also in use since the 16th century. The name *Switzer* is from the Alemannic *Schwiizer*, in origin an inhabitant of *Schwyz* and its associated territory, one of the Waldstätten cantons which formed the nucleus of the Old Swiss Confederacy. The toponym itself is first attested in 972, as Old High German *Suittes*, ultimately perhaps related to *suedan* "to burn", referring to the area of forest that was burned and cleared to build.[17] The name was extended to the area dominated by the canton, and after the Swabian War of 1499 gradually came to be used for the entire Confederation.[18] [19]

The Swiss German name of the country, *Schwiiz*, is homophonous to that of the canton and the settlement, but distinguished by the use of the definite article (*d'Schwiiz* for the Confederation,[20] but simply *Schwiiz* for the canton and the town).[21]

The New Latin name *Confederatio Helvetica* was introduced gradually after the formation of the federal state in 1848, harking back to the Napoleonic Helvetic Republic, appearing on coins from 1879, inscribed on the Federal Palace in 1902 and after 1948 used in the official seal.[22] It is derived from the name of the *Helvetii*, a Gaulish tribe living on the Swiss plateau before the Roman era. *Helvetia* appears as a national personification of the Swiss confederacy in the 17th century, with a 1672 play by Johann Caspar Weissenbach.[23]

History

Switzerland has existed as a state in its present form since the adoption of the Swiss Federal Constitution in 1848. The precursors of Switzerland established a protective alliance at the end of the 13th century (1291), forming a loose confederation of states which persisted for centuries.

Early history

The oldest traces of hominid existence in Switzerland date back about 150,000 years.[24] The oldest known farming settlements in Switzerland, which were found at Gächlingen, have been dated to around 5300 BC.[24]

Founded in 44 BC, Augusta Raurica was the first Roman settlement on the Rhine and is now amongst the most important archaeological sites in Switzerland.[25]

The earliest known cultural tribes of the area were members of the Hallstatt and La Tène cultures, named after the archaeological site of La Tène on the north side of Lake Neuchâtel. La Tène culture developed and flourished during the late Iron Age from around 450 BC,[24] possibly under some influence from the Greek and Etruscan civilisations. One of the most important tribal groups in the Swiss region was the Helvetii. In 58 BC, at the Battle of Bibracte, Julius Caesar's armies defeated the Helvetii.[24] In 15 BC, Tiberius, who was destined to be the second Roman emperor and his brother, Drusus, conquered the Alps, integrating them into the Roman Empire. The area occupied by the Helvetii—the namesakes of the later *Confoederatio Helvetica*—first became part of Rome's Gallia Belgica province and then of its Germania Superior province, while the eastern portion of modern Switzerland was integrated into the Roman province of Raetia.

In the Early Middle Ages, from the 4th century, the western extent of modern-day Switzerland was part of the territory of the Kings of the Burgundians. The Alemanni settled the Swiss plateau in the 5th century and the valleys of the Alps in the 8th century, forming Alemannia. Modern-day Switzerland was therefore then divided between the kingdoms of Alemannia and Burgundy.[24] The entire region became part of the expanding Frankish Empire in the 6th century, following Clovis I's victory over the Alemanni at Tolbiac in 504 AD, and later Frankish domination of the Burgundians.[26] [27]

Throughout the rest of the 6th, 7th and 8th centuries the Swiss regions continued under Frankish hegemony (Merovingian and Carolingian dynasties). But after its extension under Charlemagne, the Frankish empire was divided by the Treaty of Verdun in 843.[24] The territories of present day Switzerland became divided into Middle Francia and East Francia until they were reunified under the Holy Roman Empire around 1000 AD.[24]

By 1200, the Swiss plateau comprised the dominions of the houses of Savoy, Zähringer, Habsburg and Kyburg.[24] Some regions (Uri, Schwyz, Unterwalden, later known as *Waldstätten*) were accorded the Imperial immediacy to grant the empire direct control over the mountain passes. When the Kyburg dynasty fell in 1264 AD, the Habsburgs under King Rudolph I (Holy Roman Emperor in 1273) extended their territory to the eastern Swiss plateau.[26]

Old Swiss Confederacy

Further information: Growth of the Old Swiss Confederacy, Reformation in Switzerland, and Early Modern Switzerland

The Old Swiss Confederacy was an alliance among the valley communities of the central Alps. The Confederacy facilitated management of common interests (free trade) and ensured peace on the important mountain trade routes. The Federal Charter of 1291 agreed between the rural communes of Uri, Schwyz, and Unterwalden is considered the confederacy's founding document, even though similar alliances are likely to have existed decades earlier.[28] [29]

Federal charter of 1291

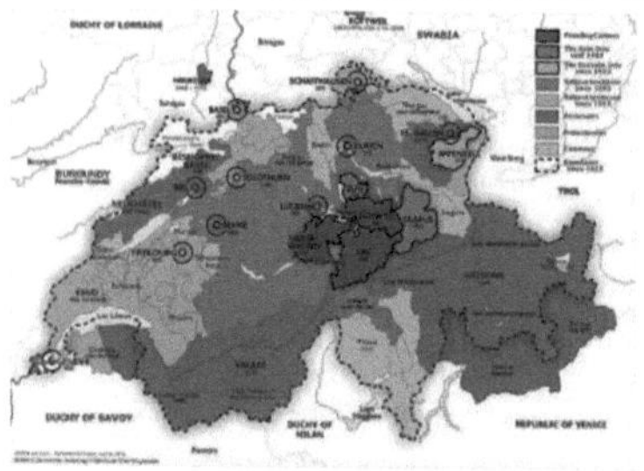

The Old Swiss Confederacy from 1291 (dark grey) to the sixteenth century (light grey) and its associates (blue)

By 1353, the three original cantons had joined with the cantons of Glarus and Zug and the Lucerne, Zurich and Bern city states to form the "Old Confederacy" of eight states that existed until the end of the 15th century. The expansion led to increased power and wealth for the federation.[29] By 1460, the confederates controlled most of the territory south and west of the Rhine to the Alps and the Jura mountains, particularly after victories against the Habsburgs (Battle of Sempach, Battle of Näfels), over Charles the Bold of Burgundy during the 1470s, and the success of the Swiss mercenaries. The Swiss victory in the Swabian War against the Swabian League of Emperor Maximilian I in 1499 amounted to *de facto* independence within the Holy Roman Empire.[29]

The Old Swiss Confederacy had acquired a reputation of invincibility during these earlier wars, but expansion of the federation suffered a setback in 1515 with the Swiss defeat in the Battle of Marignano. This ended the so-called "heroic" epoch of Swiss history.[29] The success of Zwingli's Reformation in some cantons led to inter-cantonal religious conflicts in 1529 and 1531 (Wars of Kappel). It was not until more than one hundred years after these internal wars that, in 1648, under the Peace of Westphalia, European countries recognized Switzerland's independence from the Holy Roman Empire and its neutrality.[26] [27]

During the Early Modern period of Swiss history, the growing authoritarianism of the patriciate families combined with a financial crisis in the wake of the Thirty Years' War led to the Swiss peasant war of 1653. In the background to this struggle, the conflict between Catholic and Protestant cantons persisted, erupting in further violence at the Battles of Villmergen in 1656 and 1712.[29]

Napoleonic era

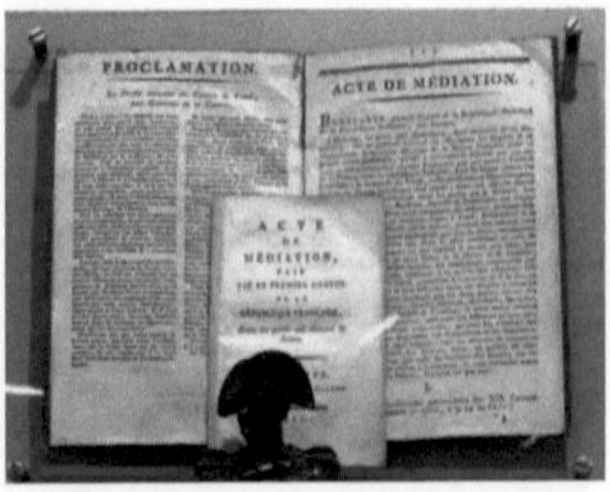
The Act of Mediation was Napoleon's attempt at a compromise between the Ancien Régime and a Republic.

In 1798, the revolutionary French government conquered Switzerland and imposed a new unified constitution.[29] This centralised the government of the country and effectively abolished the cantons and Mülhausen and Valtellina valley separated from Switzerland. The new regime, known as the Helvetic Republic, was highly unpopular. It had been imposed by a foreign invading army and destroyed centuries of tradition, making Switzerland nothing more than a French satellite state. The fierce French suppression of the Nidwalden Revolt in September 1798 was an example of the oppressive presence of the French Army and the local population's resistance to the occupation.

When war broke out between France and its rivals, Russian and Austrian forces invaded Switzerland. The Swiss refused to fight alongside the French in the name of the Helvetic Republic. In 1803 Napoleon organised a meeting of the leading Swiss politicians from both sides in Paris. The result was the Act of Mediation which largely restored Swiss autonomy and introduced a Confederation of 19 cantons.[29] Henceforth much of Swiss politics would concern balancing the cantons' tradition of self-rule with the need for a central government.

In 1815 the Congress of Vienna fully re-established Swiss independence and the European powers agreed to permanently recognise Swiss neutrality.[26] [27] [29] Swiss troops still served foreign governments until 1860 when they fought in the Siege of Gaeta. The treaty also allowed Switzerland to increase its territory, with the admission of the cantons of Valais, Neuchâtel and Geneva. Switzerland's borders have not changed since.[30]

Federal state

The first Federal Palace in Bern (1857). One of the three cantons presiding over the Tagsatzung (former legislative and executive council), Bern was chosen as the federal capital in 1848, mainly because of its closeness to the French speaking area.[31]

The restoration of the power to the patriciate was only temporary. After a period of unrest with repeated violent clashes such as the Züriputsch of 1839, civil war broke out in 1847 when some of the Catholic cantons tried to set up a separate alliance (the *Sonderbundskrieg*).[29] The war lasted for less than a month, causing fewer than 100 casualties, most of which were through friendly fire. However minor the Sonderbundskrieg seems to be when compared with other European riots and wars in the 19th century, it nevertheless had a major impact on both the psychology and the society of the Swiss and of Switzerland.

The war convinced most Swiss of the need for unity and strength towards its European neighbours. Swiss people from all strata of society, whether Catholic, Protestant, or from the liberal or conservative current, realised that the cantons would profit more if their economic and religious interests were merged.

Thus, while the rest of Europe was plagued by revolutionary uprisings, the Swiss drew up a constitution which provided for a federal layout, much of it inspired by the American example. This constitution provided for a central authority while leaving the cantons the right to self-government on local issues. Giving credit to those who favoured the power of the cantons (the Sonderbund Kantone), the national assembly was divided between an upper house (the Swiss Council of States, 2 representatives per canton) and a lower house (the National Council of Switzerland, representatives elected from across the country). Referenda were made mandatory for any amendment of this constitution.[27]

Inauguration in 1882 of the Gotthard Rail Tunnel connecting the southern canton of Ticino, the longest in the world at the time.[32]

A system of single weights and measures was introduced and in 1850 the Swiss franc became the Swiss single currency. Article 11 of the constitution forbade sending troops to serve abroad, though the Swiss were still obliged to serve Francis II of the Two Sicilies with Swiss Guards present at the Siege of Gaeta in 1860, marking the end of foreign service.

An important clause of the constitution was that it could be re-written completely if this was deemed necessary, thus enabling it to evolve as a whole rather than being modified one amendment at a time.[33]

This need soon proved itself when the rise in population and the Industrial Revolution that followed led to calls to modify the constitution accordingly. An early draft was rejected by the population in 1872 but modifications led to its acceptance in 1874.[29] It introduced the facultative referendum for laws at the federal level. It also established federal responsibility for defense, trade, and legal matters.

In 1891, the constitution was revised with unusually strong elements of direct democracy, which remain unique even today.[29]

Modern history

Switzerland was not invaded during either of the world wars. During World War I, Switzerland was home to Vladimir Illych Ulyanov (Lenin) and he remained there until 1917.[34] Swiss neutrality was seriously questioned by the Grimm-Hoffmann Affair in 1917, but it was short-lived. In 1920, Switzerland joined the League of Nations, which was based in Geneva, on the condition that it was exempt from any military requirements.

General Ulrich Wille, Commander-in-Chief of the Swiss Army during World War I

During World War II, detailed invasion plans were drawn up by the Germans,[35] but Switzerland was never attacked.[29] Switzerland was able to remain independent through a combination of military deterrence, concessions to Germany, and good fortune as larger events during the war delayed an invasion.[27] [36] Under General Henri Guisan, a massive mobilisation of militia forces was ordered. The Swiss military strategy was changed from one of static defence at the borders to protect the economic heartland, to one of organised long-term attrition and withdrawal to strong, well-stockpiled positions high in the Alps known as the Reduit. Switzerland was an important base for espionage by both sides in the conflict and often mediated communications between the Axis and Allied powers.[36]

Switzerland's trade was blockaded by both the Allies and by the Axis. Economic cooperation and extension of credit to the Third Reich varied according to the perceived likelihood of invasion and the availability of other trading partners. Concessions reached a peak after a crucial rail link through Vichy France was severed in 1942, leaving Switzerland completely surrounded by the Axis. Over the course of the war, Switzerland interned over 300,000 refugees[37] and the International Red Cross, based in Geneva, played an important part during the conflict. Strict immigration and asylum policies as well as the financial relationships with Nazi Germany raised controversy, but not until the end of the 20th century.[38] Even in the 21st century, some Swiss banks and entities still refused to surrender the assets obtained from victims of Nazi persecution.

During the war, the Swiss Air Force engaged aircraft of both sides, shooting down 11 intruding Luftwaffe planes in May and June 1940, then forcing down other intruders after a change of policy following threats from Germany. Over 100 Allied bombers and their crews were interned during the war. During 1944–45, Allied bombers mistakenly bombed a few places in Switzerland, among which were the cities of Schaffhausen, Basel and Zurich.[36]

After the war, the Swiss government exported credits through the charitable fund known as the Schweizerspende and also donated to the Marshall Plan to help Europe's recovery, efforts that ultimately benefit the Swiss economy.[39]

Women were granted the right to vote in the first Swiss cantons in 1959, at the federal level in 1971[29] [40] and, after resistance, in the last canton Appenzell Innerrhoden (one of only two remaining *Landsgemeinde*) in 1990. After suffrage at the federal level, women quickly rose in political significance, with the first woman on the seven member Federal Council executive being Elisabeth Kopp, who served from 1984–1989,[29] and the first female president being Ruth Dreifuss in 1999.

Switzerland joined the Council of Europe in 1963.[27] In 1979 areas from the canton of Bern attained independence from the Bernese, forming the new canton of Jura. On 18 April 1999 the Swiss population and the cantons voted in favour of a completely revised federal constitution.[29]

In 2003, by granting the Swiss People's Party a second seat in the governing cabinet, the Parliament altered the coalition which had brought stability to Swiss politics since 1959.

In 2002 Switzerland became a full member of the United Nations, leaving the Vatican City as the last widely recognised state without full UN membership. Switzerland is a founding member of the EFTA, but is not a member of the European Economic Area. An application for membership in the European Union was sent in May 1992, but not advanced since the EEA was rejected in December 1992[29] when Switzerland was the only country to launch a referendum on the EEA. There have since been several referenda on the EU issue; due to a mixed reaction from the population the membership application has been frozen. Nonetheless, Swiss law is gradually being adjusted to conform with that of the EU, and the government has signed a number of bilateral agreements with the European Union. Switzerland, together with Liechtenstein, has been completely surrounded by the EU since Austria's membership in 1995. On 5 June 2005, Swiss voters agreed by a 55% majority to join the Schengen treaty, a result that was regarded by EU commentators as a sign of support by Switzerland, a country that is traditionally perceived as independent and reluctant to enter supranational bodies.[27]

Geography

Extending across the north and south side of the Alps, Switzerland encompasses a great diversity of landscapes and climates on a limited area of 41285 square kilometres (**unknown operator: u'strong'** sq mi).[41] The population is about 7.9 million, resulting in an average population density of around 190 people per square kilometre (485/sq mi).[41] [42] [43] The more mountainous southern half of the country is far more sparsely populated than the northern half.[41] In the largest Canton of Graubünden, lying entirely in the Alps, population density falls to 27 /km² (70 /sq mi).

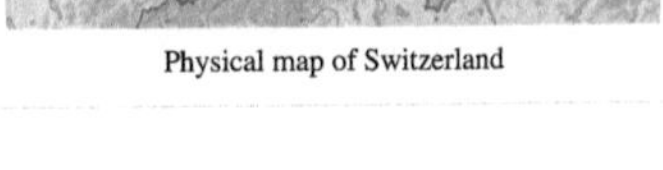

Physical map of Switzerland

Contrasted landscapes between the Matterhorn area in the high Alps, the Sanetsch region and the plateau at Lake Lucerne

Switzerland lies between latitudes 45° and 48° N, and longitudes 5° and 11° E. It contains three basic topographical areas: the Swiss Alps to the south, the Swiss plateau or middleland, and the Jura mountains on the north. The Alps are a high mountain range running across the central-south of the country, comprising about 60% of the country's total area. Among the high valleys of the Swiss Alps many glaciers are found, totalling an area of 1,063 square kilometres. From these originate the headwaters of several major rivers, such as the Rhine, Inn, Ticino and Rhone, which flow in the four cardinal directions into the whole of Europe. The hydrographic network includes several of the largest bodies of freshwater in Central and Western Europe, among which are included Lake Geneva, Lake Constance and Lake Maggiore. Switzerland has more than 1500 lakes, and contains 6% of Europe's stock of fresh water. Lakes and glaciers cover about 6% of the national territory.[41] [44] [45]

About a hundred of Switzerland's mountain peaks are close to or higher than 4000 metres (**unknown operator: u'strong'** ft).[46] At 4634 m (**unknown operator: u'strong'** ft), Monte Rosa is the highest, although the Matterhorn (4478 m/**unknown operator: u'strong'** ft) is probably the most famous. Both are located within the Pennine Alps in the canton of Valais. The section of the Bernese Alps above the deep glacial Lauterbrunnen valley, containing 72 waterfalls, is well known for the Jungfrau (4158 m/**unknown operator: u'strong'** ft) and Eiger, and the many picturesque valleys in the region. In the southeast the long Engadin Valley, encompassing the St. Moritz area in canton Graubünden, is also well known; the highest peak in the neighbouring Bernina Alps is Piz Bernina (4049 m/**unknown operator: u'strong'** ft).[47]

The more populous northern part of the country, comprising about 30% of the country's total area, is called the Middle Land. It has greater open and hilly landscapes, partly forested, partly open pastures, usually with grazing herds, or vegetables and fruit fields, but it is still hilly. There are large lakes found here and the biggest Swiss cities are in this area of the country.[47] The largest lake is Lake Geneva (also called Lac Léman in French), in western Switzerland. The Rhone River is both the main input and output of Lake Geneva.

Climate

Contrasted climates between the most glaciated area in western Eurasia (Aletsch Glacier),[48] the cold temperate Jura (Vallée de Joux) and the southern canton of Ticino (Lake Lugano)

The Swiss climate is generally temperate, but can vary greatly between the localities,[49] [50] from glacial conditions on the mountaintops to the often pleasant near Mediterranean climate at Switzerland's southern tip. Summers tend to be warm and humid at times with periodic rainfall so they are ideal for pastures and grazing. The less humid winters in the mountains may see long intervals of stable conditions for weeks, while the lower lands tend to suffer from inversion, during these periods, thus seeing no sun for weeks.

A weather phenomenon known as the föhn (with an identical effect to the chinook wind) can occur at all times of the year and is characterised by an unexpectedly warm wind, bringing air of very low relative humidity to the north of the Alps during rainfall periods on the southern face of the Alps. This works both ways across the alps but is more efficient if blowing from the south due to the steeper step for oncoming wind from the south. Valleys running south to north trigger the best effect. The driest conditions persist in all inner alpine valleys that receive less rain because arriving clouds lose a lot of their content while crossing the mountains before reaching these areas. Large alpine areas such as Graubünden remain drier than pre-alpine areas and as in the main valley of the Valais wine grapes are grown there.[51]

The wettest conditions persist in the high Alps and in the Ticino canton which has much sun yet heavy bursts of rain from time to time.[51] Precipitation tends to be spread moderately throughout the year with a peak in summer. Autumn is the driest season, winter receives less precipitation than summer, yet the weather patterns in Switzerland are not in a stable climate system and can be variable from year to year with no strict and predictable periods.

Environment

Switzerland's ecosystems can be particularly fragile, because of the many delicate valleys separated by high mountains, often forming unique ecologies. The mountainous regions themselves are also vulnerable, with a rich range of plants not found at other altitudes, and experience some pressure from visitors and grazing. The climatic, geological and topographical conditions of the alpine region make for a very fragile ecosystem that is particularly sensitive to climate change.[52] [53]

Politics

The Federal Constitution adopted in 1848 is the legal foundation of the modern federal state. It is among the oldest constitutions in the world.[55] A new Constitution was adopted in 1999, but did not introduce notable changes to the federal structure. It outlines basic and political rights of individuals and citizen participation in public affairs, divides the powers between the Confederation and the cantons and defines federal jurisdiction and authority. There are three main governing bodies on the federal level:[56] the bicameral parliament (legislative), the Federal Council (executive) and the Federal Court (judicial).

The Swiss Federal Council in 2012 with President Eveline Widmer-Schlumpf (in the middle)[54]

The Federal Palace, seat of the Federal Assembly and the Federal Council.

The Swiss Parliament consists of two houses: the Council of States which has 46 representatives (two from each canton and one from each half-canton) who are elected under a system determined by each canton, and the National Council, which consists of 200 members who are elected under a system of proportional representation, depending on the population of each canton. Members of both houses serve for 4 years. When both houses are in joint session, they are known collectively as the Federal Assembly. Through referendums, citizens may challenge any law passed by parliament and through initiatives, introduce amendments to the federal constitution, thus making Switzerland a direct democracy.[55]

The Federal Council constitutes the federal government, directs the federal administration and serves as collective Head of State. It is a collegial body of seven members, elected for a four-year mandate by the Federal Assembly which also exercises oversight over the Council. The President of the Confederation is elected by the Assembly from among the seven members, traditionally in rotation and for a one-year term; the President chairs the government and assumes representative functions. However, the president is a *primus inter pares* with no additional powers, and remains the head of a department within the administration.[55]

The Swiss government has been a coalition of the four major political parties since 1959, each party having a number of seats that roughly reflects its share of electorate and representation in the federal parliament. The classic distribution of 2 CVP/PDC, 2 SPS/PSS, 2 FDP/PRD and 1 SVP/UDC as it stood from 1959 to 2003 was known as the "magic formula". In the 2007 Federal Council elections the seven seats in the Federal Council were distributed as follows:

2 Social Democrats (SPS/PSS),

2 Liberal Democrats (FDP/PRD),

2 Swiss People's Party (SVP/UDC),[57]

1 Christian Democrats (CVP/PDC).

The function of the Federal Supreme Court is to hear appeals against rulings of cantonal or federal courts. The judges are elected by the Federal Assembly for six-year terms.[58]

Direct democracy

The Landsgemeinde is an old form of direct democracy. It is still practised in two cantons.[59]

Swiss citizens are subject to three legal jurisdictions: the commune, canton and federal levels. The 1848 federal constitution defines a system of direct democracy (sometimes called *half-direct* or representative direct democracy since it is aided by the more commonplace institutions of a parliamentary democracy). The instruments of Swiss direct democracy at the federal level, known as civic rights (*Volksrechte*, *droits civiques*), include the right to submit a *constitutional initiative* and a *referendum*, both of which may overturn parliamentary decisions.[55] [60]

By calling a federal *referendum* a group of citizens may challenge a law that has been passed by Parliament, if they can gather 50,000 signatures against the law within 100 days. If so, a national vote is scheduled where voters decide by a simple majority whether to accept or reject the law. Eight cantons together can also call a referendum on a federal law.[55]

Similarly, the federal *constitutional initiative* allows citizens to put a constitutional amendment to a national vote, if they can get 100,000 voters to sign the proposed amendment within 18 months.[61] Parliament can supplement the proposed amendment with a counter-proposal, with voters having to indicate a preference on the ballot in case both proposals are accepted. Constitutional amendments, whether introduced by initiative or in Parliament, must be accepted by a double majority of both the national popular vote and a majority of the cantonal popular votes.[62] [63] [64] [65]

Administrative divisions

The Swiss Confederation consists of 26 cantons:[55]

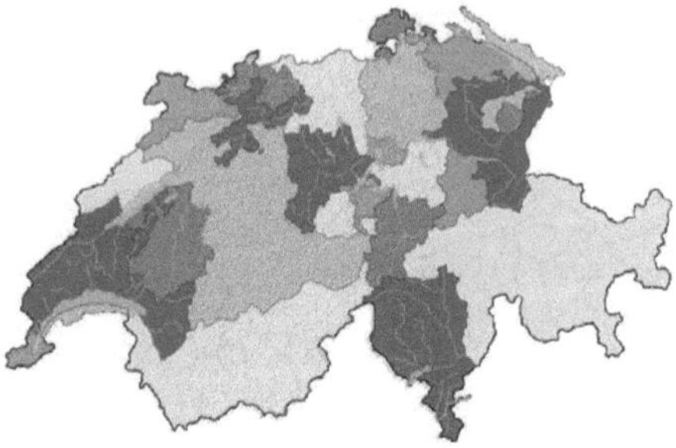

Valais

Ticino

Graubünden (Grisons)

Geneva

Vaud

Neuchâtel

Jura

Bern

Thurgau

Zurich

Aargau

Lucerne

Solothurn

Basel-Land

Schaffhausen

Uri

Schwyz

Glarus

St. Gallen

AI

AR

Obwalden

Nidwalden

Zug

Fribourg

Basel-Stadt

France

Italy

Liecht.

Austria

Germany

Canton	Capital	Canton	Capital
Aargau	Aarau	*Nidwalden	Stans
*Appenzell Ausserrhoden	Herisau	*Obwalden	Sarnen
*Appenzell Innerrhoden	Appenzell	Schaffhausen	Schaffhausen
*Basel-Landschaft	Liestal	Schwyz	Schwyz
*Basel-Stadt	Basel	Solothurn	Solothurn
Bern	Bern	St. Gallen	St. Gallen
Fribourg	Fribourg	Thurgau	Frauenfeld
Geneva	Geneva	Ticino	Bellinzona
Glarus	Glarus	Uri	Altdorf
Graubünden	Chur	Valais	Sion
Jura	Delémont	Vaud	Lausanne
Lucerne	Lucerne	Zug	Zug
Neuchâtel	Neuchâtel	Zurich	Zurich

*These half-cantons are represented by one councillor (instead of two) in the Council of States.

The cantons have a permanent constitutional status and, in comparison with the situation in other countries, a high degree of independence. Under the Federal Constitution, all 26 cantons are equal in status. Each canton has its own constitution, and its own parliament, government and courts.[59] However, there are considerable differences between the individual cantons, most particularly in terms of population and geographical area. Their populations vary between 15,000 (Appenzell Innerrhoden) and 1,253,500 (Zurich), and their area between 37 km^2 (**unknown operator: u'strong'** sq mi) (Basel-Stadt) and 7105 km^2 (**unknown operator: u'strong'** sq mi) (Graubünden). The Cantons comprise a total of 2,889 municipalities. Within Switzerland there are two enclaves: Büsingen belongs to Germany, Campione d'Italia belongs to Italy.[66]

In a referendum held in the Austrian state of Vorarlberg on 11 May 1919 over 80% of those voting supported a proposal that the state should join the Swiss Confederation. However, this was prevented by the opposition of the Austrian Government, the Allies, Swiss liberals and non German-speaking Swiss.[67] [68]

Foreign relations and international institutions

Traditionally, Switzerland avoids alliances that might entail military, political, or direct economic action and had been neutral since the end of its expansion in 1515. Its policy of neutrality was internationally recognised at the Congress of Vienna in 1815.[69] [70] Only in 2002 did Switzerland become a full member of the United Nations[69] and it was the first state to join it by referendum. Switzerland maintains diplomatic relations with almost all countries and historically has served as an intermediary between other states.[69] Switzerland is not a member of the European Union; the Swiss people have consistently rejected membership since the early 1990s.[69]

The Palace of Nations, the European headquarters of the United Nations in Geneva

The monochromatically reversed Swiss flag became the symbol of the Red Cross Movement,[40] founded in 1863 by Henri Dunant.[71]

An unusual number of international institutions have their seats in Switzerland, in part because of its policy of neutrality. Geneva is the birth place of the Red Cross and Red Crescent Movement and the Geneva Conventions and, since 2006, hosts the United Nations Human Rights Council. Even though Switzerland is one of the most recent countries to have joined the United Nations, the Palace of Nations in Geneva is the second biggest centre for the United Nations after New York, and Switzerland was a founding member and home to the League of Nations.

Apart from the United Nations headquarters, the Swiss Confederation is host to many UN agencies, like the World Health Organization (WHO), the International Labour Organization (ILO), the International Telecommunication Union (ITU), the United Nations High Commissioner for Refugees (UNHCR) and about 200 other international organisations, including the World Trade Organization.[69] The annual meetings of the World Economic Forum in Davos bring together top international business and political leaders from Switzerland and foreign countries to discuss important issues facing the world, including health and the environment.

Furthermore, many sport federations and organisations are located throughout the country, such as the International Basketball Federation, in Geneva, the UEFA (Union of European Football Associations), in Nyon, the FIFA (International Federation of Association Football) and the International Ice Hockey Federation, in Zurich, the International Cycling Union, in Aigle, and the International Olympic Committee, in Lausanne.[72]

Military

A Swiss Air Force F/A-18 Hornet in flight over the Alps

The Swiss Armed Forces, including the Land Forces and the Air Force, are composed of conscripts: professional soldiers constitute only about 5 percent of the military personnel, and all the rest are conscript male citizens aged from 20 to 34 (in special cases up to 50) years. Being a landlocked country, Switzerland has no navy; however, on lakes bordering neighbouring countries, armed military patrol boats are used. Swiss citizens are prohibited from serving in foreign armies, with the exception of the Swiss Guards of the Vatican.

The structure of the Swiss militia system stipulates that the soldiers keep their Army issued equipment, including all personal weapons, at home. Some organizations and political parties find this practice controversial[73] but mainstream Swiss opinion is in favour of the system. Compulsory military service concerns all male Swiss citizens; women can serve voluntarily. Men usually receive military conscription orders for training at the age of 19. About two thirds of the young Swiss are found suited for service; for those found unsuited, various forms of alternative service exist.[74] Annually, approximately 20,000 persons are trained in recruit centres for a duration from 18 to 21 weeks. The reform "Army XXI" was adopted by popular vote in 2003, it replaced the previous model "Army 95", reducing the effectives from 400,000 to about 200,000. Of those, 120,000 are active in periodic Army training and 80,000 are non-training reserves.[75]

MOWAG Eagle armoured vehicles in a military parade

Overall, three general mobilisations have been declared to ensure the integrity and neutrality of Switzerland. The first one was held on the occasion of the Franco-Prussian War of 1870–71. The second one was decided in response to the outbreak of the First World War in August 1914. The third mobilisation of the army took place on September 1939 in response to the German attack on Poland; Henri Guisan was elected as the General-in-Chief.

Because of its neutrality policy, the Swiss army does not currently take part in armed conflicts in other countries, but is part of some peacekeeping missions around the world. Since 2000 the armed forces department has also maintained the Onyx intelligence gathering system to monitor satellite communications.[76]

Following the end of the Cold War there have been a number of attempts to curb military activity or even abolish the armed forces altogether. A notable referendum on the subject, launched by an anti-militarist group, was held on 26 November 1989. It was defeated with about two thirds of the voters against the proposal.[77] [78] A similar referendum, called for before, but held shortly after, the September 11 attacks in the US, was defeated by over 78% of voters.[79]

Economy

The Omega *Speedmaster* worn on the moon during the Apollo missions. In terms of value, Switzerland is responsible for half of the world production of watches.[40] [80]

Switzerland has a stable, prosperous and high-tech economy. In 2011, it was ranked as being the wealthiest country in the world in per capita terms (with 'wealth' being defined to include both financial and non-financial assets).[81] [82] It has the world's nineteenth largest economy by nominal GDP and the thirty-sixth largest by purchasing power parity. It is the twentieth largest exporter and eighteenth largest importer of goods. Switzerland has the highest European rating in the Index of Economic Freedom 2010, while also providing large coverage through public services.[83] The nominal per capita GDP is higher than those of the larger Western and Central European economies and Japan.[84] If adjusted for purchasing power parity, Switzerland ranks 8th in the world in terms of GDP per capita, according to the World Bank and IMF (ranked 15th according to the CIA Worldfactbook[84]).

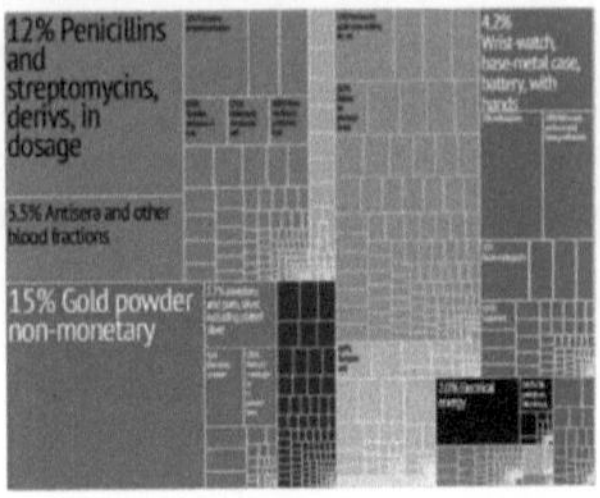

Graphical depiction of Switzerland's product exports in 28 color coded categories.

The World Economic Forum's Global Competitiveness Report currently ranks Switzerland's economy as the most competitive in the world,[85] while ranked by the European Union as Europe's most innovative country.[86] For much of the 20th century, Switzerland was the wealthiest country in Europe by a considerable margin (by GDP - per capita).[87] In 2005 the median household income in Switzerland was an estimated 95,000 CHF, the equivalent of roughly 100,000 USD (as of December 2010) in nominal terms. Switzerland also has one of the world's largest account balances as a percentage of GDP.

The Greater Zurich Area, home to 1.5 million inhabitants and 150,000 companies, is one of the most important economic centers in the world.[88]

Switzerland is home to several large multinational corporations. The largest Swiss companies by revenue are Glencore, Nestlé, Novartis, Hoffmann-La Roche, ABB and Adecco.[89] Also notable are UBS AG, Zurich Financial Services, Credit Suisse, Swiss Re, and The Swatch Group. Switzerland is ranked as having one of the most powerful economies in the world.[87]

Switzerland's most important economic sector is manufacturing. Manufacturing consists largely of the production of specialist chemicals, health and pharmaceutical goods, scientific and precision measuring instruments and musical instruments. The largest exported goods are chemicals (34% of exported goods), machines/electronics (20.9%), and precision instruments/watches (16.9%).[90] Exported services amount to a third of exported goods.[90] The services sector - especially banking and insurance, tourism, and international organisations - is another important industry for Switzerland.

Around 3.8 million people work in Switzerland. Switzerland has a more flexible job market than neighboring countries and the unemployment rate is very low. Unemployment rate increased from a low of 1.7% in June 2000 to a peak of 4.4%, as of December 2009.[91] Population growth from net immigration is quite high, at 0.52% of population in 2004.[90] Foreign citizen population is 21.8% as of 2004,[90] about the same as in Australia. GDP per hour worked is the world's 17th highest, at 27.44 international dollars in 2006.

The Engadin Valley. Tourism constitutes an important revenue for the less industrialised alpine regions.

Switzerland has an overwhelmingly private sector economy and low tax rates by the Western World standards; overall taxation is one of the smallest of developed countries. Switzerland is an easy place to do business; Switzerland currently ranks 27th of 178 countries in the Ease of Doing Business Index. The slow growth Switzerland experienced in the 1990s and the early 2000s has brought greater support for economic reforms and harmonisation with the European Union.[92] [93] According to Credit Suisse, only about 37% of residents own their own homes, one of the lowest rates of home ownership in Europe. Housing and food price levels were 171% and 145% of the EU-25 index in 2007, compared to 113% and 104% in Germany.[90]

Agricultural protectionism—a rare exception to Switzerland's free trade policies—has contributed to high food prices. Product market liberalisation is lagging behind many EU countries according to the OECD.[92] Nevertheless, domestic purchasing power is one of the best in the world.[94] [95] [96] Apart from agriculture, economic and trade barriers between the European Union and Switzerland are minimal and Switzerland has free trade agreements worldwide. Switzerland is a member of the European Free Trade Association (EFTA).

Education and science

Education in Switzerland is very diverse because the constitution of Switzerland delegates the authority for the school system to the cantons.[97] There are both public and private schools, including many private international schools. The minimum age for primary school is about six years in all cantons, but most cantons provide a free "children's school" starting at four or five years old.[97] Primary school continues until grade four, five or six, depending on the school. Traditionally, the first foreign language in school was always one of the other national languages, although recently (2000) English was introduced first in a few cantons.[97]

Some of the Swiss scientists who played a key role in their discipline (clockwise):
Leonhard Euler (mathematics)
Louis Agassiz (glaciology)
Auguste Piccard (aeronautics)
Albert Einstein (physics)

At the end of primary school (or at the beginning of secondary school), pupils are separated according to their capacities in several (often three) sections. The fastest learners are taught advanced classes to be prepared for further studies and the matura,[97] while students who assimilate a little bit more slowly receive an education more adapted to their needs.

The campus of the ETH Zurich. The institution is usually ranked the top university in continental Europe.[98] [99]

There are 12 universities in Switzerland, ten of which are maintained at cantonal level and usually offer a range of non-technical subjects. The first university in Switzerland was founded in 1460 in Basel (with a faculty of medicine) and has a tradition of chemical and medical research in Switzerland. The biggest university in Switzerland is the University of Zurich with nearly 25,000 students. The two institutes sponsored by the federal government are the ETHZ in Zurich (founded 1855) and the EPFL in Lausanne (founded 1969 as such, formerly an institute associated with the University of Lausanne) which both have an excellent international reputation.[100] [101]

In addition there are various Universities of Applied Sciences. In business and management studies, University of St. Gallen, (HSG) and International Institute for Management Development (IMD) are the leaders within the country and highly regarded internationally. Switzerland has the second highest rate of foreign students in tertiary education, after Australia.[102]

As might befit a country that plays home to innumerable international organizations, the Graduate Institute of International and Development Studies, located in Geneva, is not only continental Europe's oldest graduate school of international and development studies, but also widely believed to be one of its most prestigious.[103] [104]

Many Nobel prizes were awarded to Swiss scientists, for example to the world-famous physicist Albert Einstein[105] in the field of physics who developed his Special relativity while working in Bern. More recently Vladimir Prelog, Heinrich Rohrer, Richard Ernst, Edmond Fischer, Rolf Zinkernagel and Kurt Wüthrich received Nobel prizes in the sciences. In total, 113 Nobel Prize winners in all fields stand in relation to Switzerland[106] [107] and the Nobel Peace Prize was awarded 9 times to organisations residing in Switzerland.[108]

Geneva and the nearby French department of Ain co-host the world's largest laboratory, CERN,[110] [111] dedicated to particle physics research. Another important research center is the Paul Scherrer Institute. Notable inventions include lysergic acid diethylamide (LSD), the scanning tunneling microscope (Nobel prize) and Velcro. Some technologies enabled the exploration of new worlds such as the pressurized balloon of Auguste Piccard and the Bathyscaphe which permitted Jacques Piccard to reach the deepest point of the world's oceans.

The LHC tunnel. CERN is the world's largest laboratory and also the birthplace of the World Wide Web.[109]

Switzerland Space Agency, the Swiss Space Office, has been involved in various space technologies and programs. In addition it was one of the 10 founders of the European Space Agency in 1975 and is the seventh largest contributor to the ESA budget. In the private sector, several companies are implicated in the space industry such as Oerlikon Space[112] or Maxon Motors[113] who provide spacecraft structures.

Switzerland and the European Union

Switzerland voted against membership in the European Economic Area in a referendum in December 1992 and has since maintained and developed its relationships with the European Union (EU) and European countries through bilateral agreements. In March 2001, the Swiss people refused in a popular vote to start accession negotiations with the EU.[114] In recent years, the Swiss have brought their economic practices largely into conformity with those of the EU in many ways, in an effort to enhance their international competitiveness. The economy has been growing most recently at around 3% per year. Full EU membership is a long-term objective of some in the Swiss government, but there is considerable popular sentiment against this supported by the conservative SVP party. The western French-speaking areas and the urban regions of the rest of the country tend to be more pro-EU, however with far from any significant share of the population.[115] [116]

The government has established an Integration Office under the Department of Foreign Affairs and the Department of Economic Affairs. To minimise the negative consequences of Switzerland's isolation from the rest of Europe, Bern and Brussels signed seven bilateral agreements to further liberalise trade ties. These agreements were signed in 1999 and took effect in 2001. This first series of bilateral agreements included the free movement of persons. A second series covering nine areas was signed in 2004 and has since been ratified. The second series includes the Schengen Treaty and the Dublin Convention. They continue to discuss further areas for cooperation.

In 2006, Switzerland approved a billion francs supportive investment in the poorer Southern and Central European countries in support of cooperation and positive ties to the EU as a whole. A further referendum will be needed to approve 300 million francs to support Romania and Bulgaria and their recent admission. The Swiss have also been under EU and sometimes international pressure to reduce banking secrecy and to raise tax rates to parity with the EU. Preparatory discussions are being opened in four new areas: opening up the electricity market, participation in the European GNSS project Galileo, cooperating with the European centre for disease prevention and recognising certificates of origin for food products.[117]

On 27 November 2008, the interior and justice ministers of European Union in Brussels announced Switzerland's accession to the Schengen passport-free zone from 12 December 2008. The land border checkpoints will remain in place only for goods movements, but should not run controls on people, though people entering the country had their passports checked until 29 March 2009 if they originated from a Schengen nation.[118]

Energy, infrastructure, and environment

The Gösgen Nuclear Power Plant is one of four nuclear power plants in Switzerland.

Electricity generated in Switzerland is 56% from hydroelectricity and 39% from nuclear power, with 5% of the electricity generated from conventional power sources resulting in a nearly CO_2-free electricity-generating network. On 18 May 2003, two anti-nuclear initiatives were turned down: *Moratorium Plus*, aimed at forbidding the building of new nuclear power plants (41.6% supported and 58.4% opposed),[119] and Electricity Without Nuclear (33.7% supported and 66.3% opposed).[120]

The former ten-year moratorium on the construction of new nuclear power plants was the result of a citizens' initiative voted on in 1990 which had passed with 54.5% Yes vs. 45.5% No votes. A new nuclear plant in the Canton of Bern is presently planned. The Swiss Federal Office of Energy (SFOE) is the office responsible for all questions relating to energy supply and energy use within the Federal Department of Environment, Transport, Energy and Communications (DETEC). The agency is supporting the 2000-watt society initiative to cut the nation's energy use by more than half by the year 2050.[121]

On 25 May 2011 the Swiss government announced that it plans to end its use of nuclear energy in the next 2 or 3 decades. "The government has voted for a phaseout because we want to ensure a secure and autonomous supply of energy", Energy Minister Doris Leuthard said that day at a press conference in Bern. "Fukushima showed that the risk of nuclear power is too high, which in turn has also increased the costs of this energy form." The first reactor would reportedly be taken offline in 2019 and the last one in 2034. Parliament will discuss the plan in June 2011, and there could be a referendum as well.[122]

Entrance of the new Lötschberg Base Tunnel, the third-longest railway tunnel in the world, under the old Lötschberg railway line. It is the first completed tunnel of the greater project AlpTransit.

The most dense rail network in Europe[40] of 5063 km (**unknown operator: u'strong'** mi) carries over 350 million passengers annually.[123] In 2007, each Swiss citizen travelled on average 2103 km (**unknown operator: u'strong'** mi) by rail, which makes them the keenest rail users.[124] The network is administered mainly by the Federal Railways, except in Graubünden, where the 366 km (**unknown operator: u'strong'** mi) narrow gauge railway is operated by the Rhaetian Railways and includes some World Heritage lines.[125] The building of new railway base tunnels through the Alps is under way to reduce the time of travel between north and south through the AlpTransit project.

Swiss private-public managed road network is funded by road tolls and vehicle taxes. The Swiss autobahn/autoroute system requires the purchase of a vignette (toll sticker)—which costs 40 Swiss francs—for one calendar year in order to use its roadways, for both passenger cars and trucks. The Swiss autobahn/autoroute network has a total length of 1638 km (**unknown operator: u'strong'** mi) (as of 2000) and has, by an area of 41290 km^2 (**unknown operator: u'strong'** sq mi), also one of the highest motorway densities in the world. Zurich Airport is Switzerland's largest international flight gateway, which handled 22.8 million passengers in 2010. The other international airports are Geneva Airport (11.8 million passengers), EuroAirport Basel-Mulhouse-Freiburg which is located in France, Bern Airport, Lugano Airport, St. Gallen-Altenrhein Airport and Sion Airport.

Switzerland has one of the best environmental records among nations in the developed world;[126] it was one of the countries to sign the Kyoto Protocol in 1998 and ratified it in 2003. With Mexico and the Republic of Korea it forms the Environmental Integrity Group (EIG).[127] The country is heavily active in recycling and anti-littering regulations and is one of the top recyclers in the world, with 66% to 96% of recyclable materials being recycled, depending on the area of the country.[128]

In many places in Switzerland, household garbage disposal is charged for. Garbage (except dangerous items, batteries etc.) is only collected if it is in bags which either have a payment sticker attached, or in official bags with the surcharge paid at the time of purchase.[129] This gives a financial incentive to recycle as much as possible, since recycling is free.[130] Illegal disposal of garbage is not tolerated but usually the enforcement of such laws is limited to violations that involve the unlawful disposal of larger volumes at traffic intersections and public areas. Fines for not paying the disposal fee range from CHF 200–500.[131]

Switzerland also has internationally the most efficient system to recycle old newspapers and cardboard materials. Publicly organized collection by volunteers and economical railway transport logistics started as early as 1865 under the leadership of the notable industrialist Hans Caspar Escher (Escher Wyss AG) when the first modern Swiss paper manufacturing plant was built in Biberist.[132]

Demographics

Further information: List of Swiss people

Switzerland lies at the crossroads of several major European cultures that have heavily influenced the country's languages and culture. Switzerland has four official languages: German (63.7% total population share, with foreign residents; 72.5% of residents with Swiss citizenship, in 2000) in the north, east and centre of the country; French (20.4%; 21.0%) to the west; Italian (6.5%; 4.3%) in the south.[133] Romansh (0.5%; 0.6%), a Romance language spoken locally in the southeastern trilingual canton of Graubünden, is designated by the Federal Constitution as a national language along with German, French and Italian (Article 4 of the Constitution), and as official language if the authorities communicate with persons of Romansh language (Article 70), but federal laws and other official acts do not need to be decreed in this language. The federal government is obliged to communicate in the official languages, and in the federal parliament simultaneous translation is provided from and into German, French and Italian.[134]

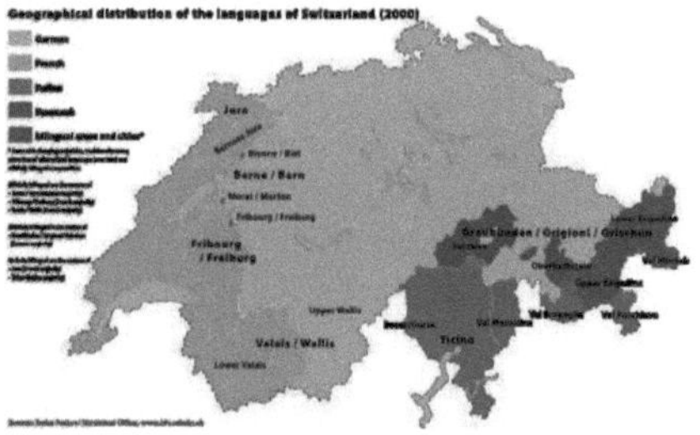

Official languages in Switzerland:Swiss Federal Statistical Office. "Languages and religions – Data, indicators". . Retrieved 9 October 2007. The first number refers to the share of languages within total population. The second refers to the Swiss citizens only. German (63.7%; 72.5%) French (20.4%; 21.0%) Italian (6.5%; 4.3%) Romansh (0.5%; 0.6%)

The German spoken in Switzerland is predominantly a group of Alemannic dialects collectively known as Swiss German, but written communication typically use Swiss Standard German, whilst the majority of radio and TV broadcast is now in Swiss German as well. Similarly, there are some dialects of Franco-Provençal in rural communities in the French speaking part, known as "Suisse romande", called Vaudois, Gruérien, Jurassien, Empro, Fribourgeois, Neuchâtelois, and in the Italian speaking area, Ticinese (a dialect of Lombard). Moreover, the official languages (German, French and Italian) borrow some terms not understood outside of Switzerland, i.e. terms from other languages (German *Billette*[135] from French), from similar term in another language (Italian *azione* used not as *act* but as *discount* from German *Aktion*). Learning one of the other national languages at school is obligatory for all Swiss, so many Swiss are supposed to be at least bilingual, especially those belonging to minorities.[136]

Resident foreigners and temporary foreign workers make up about 22% of the population.[137] Most of these (60%) are from European Union or EFTA countries.[138] Italians are the largest single group of foreigners with 17.3% of total foreign population. They are followed by Germans (13.2%), immigrants from Serbia and Montenegro (11.5%)

as well as Portugal (11.3%).[138] Immigrants from Sri Lanka, most of them former Tamil refugees, are the largest group among people of Asian origin.[139] In the 2000s, domestic and international institutions have expressed concern about what they perceive as an increase of xenophobia, particularly in some political campaignings. However, the high proportion of foreign citizens in the country, as well as the generally unproblematic integration of foreigners, underlines Switzerland's openness.[140]

Health

Swiss citizens are required to buy universal health insurance from private insurance companies, which in turn are required to accept every applicant. This system permits access to a broad range of modern medical services. The healthcare system compares well with other European countries and patients are largely satisfied with it. In 2006 life expectancy at birth was 79 years for men and 84 years for women.[141] It is among the highest in the world.[142] [143] However, spending on health is particularly high, with 11.5% of GDP (2003) and, from 1990, a steady increase is observed, reflecting the high prices of the services provided.[144] With aging populations and new healthcare technologies, health spending will likely continue to rise.[144]

Urbanization

Between two thirds and three quarters of the population live in urban areas.[145] [146] Switzerland has gone from a largely rural country to an urban one in just 70 years. Since 1935 urban development has claimed as much of the Swiss landscape as it did during the previous 2,000 years. This urban sprawl does not only affect the plateau but also the Jura and the Alpine foothills[147] and there are growing concerns about land use.[148] However from the beginning of the 21st century, the population growth in urban areas is higher than in the countryside.[146]

Urbanisation in the Rhone Valley (outskirts of Sion)

Switzerland has a dense network of cities, where large, medium and small cities are complementary.[146] The plateau is very densely populated with about 450 people per km^2 and the landscape continually shows signs of man's presence.[149] The weight of the largest metropolitan areas, which are Zurich, Geneva–Lausanne, Basel and Bern tend to increase.[146] In international comparison the importance of these urban areas is stronger than their number of inhabitants suggests.[146] In addition the two main centers of Zurich and Geneva are recognized for their particular great quality of life.[150]

Religion

Switzerland has no official state religion, though most of the cantons (except Geneva and Neuchâtel) recognize official churches, which are either the Catholic Church or the (Protestant) Swiss Reformed Church. These churches, and in some cantons also the Old Catholic Church and Jewish congregations, are financed by official taxation of adherents.[151]

The reformed church of Glarus

Christianity is the predominant religion of Switzerland, divided between the Catholic Church (41.8% of the population) and various Protestant denominations (35.3%). Geneva converted to Protestantism in 1536, just before John Calvin arrived there. Immigration has brought Islam (4.26%) and Eastern Orthodoxy (1.8%) as sizeable minority religions.[152] Other Christian minority communities include Neo-Pietism (0.44%), Pentecostalism (0.28%, mostly incorporated in the Schweizer Pfingstmission), Methodism (0.13%), the New Apostolic Church (0.45%), Jehovah's Witnesses (0.28%), other Protestant denominations (0.20%), the Old Catholic Church (0.18%), other Christian denominations (0.20%). Minor non-Christian minority groups are Hinduism (0.38%), Buddhism (0.29%), Judaism (0.25%) and "other religions" (0.11%). 4.3% did not make a statement.[153] The 2005 Eurobarometer poll[154] found 48% to be theist, 39% expressing belief in "a spirit or life force", 9% atheist and 4% agnostic. Greeley (2003) found that 27% of the population does not believe in a God.[155]

The country is historically about evenly balanced between Catholic and Protestant, with a complex patchwork of majorities over most of the country. One canton, Appenzell, was officially divided into Catholic and Protestant sections in 1597.[156] The larger cities (Bern, Geneva, Zurich and Basel) are predominantly Protestant. Central Switzerland, as well as Ticino, is traditionally Catholic. The Swiss Constitution of 1848, under the recent impression of the clashes of Catholic vs. Protestant cantons that culminated in the Sonderbundskrieg, consciously defines a consociational state, allowing the peaceful co-existence of Catholics and Protestants. A 1980 initiative calling for the complete separation of church and state was rejected by 78.9% of the voters.[157]

Culture

Alphorn concert in Vals

Three of Europe's major languages are official in Switzerland. Swiss culture is characterised by diversity, which is reflected in a wide range of traditional customs.[158] A region may be in some ways strongly culturally connected to the neighbouring country that shares its language, the country itself being rooted in western European culture.[159] The linguistically isolated Romansh culture in Graubünden in eastern Switzerland constitutes an exception, it survives only in the upper valleys of the Rhine and the Inn and strives to maintain its rare linguistic tradition.

Switzerland is home to many notable contributors to literature, art, architecture, music and sciences. In addition the country attracted a number of creative persons during time of unrest or war in Europe.[160] Some 1000 museums are distributed through the country; the number has more than tripled since 1950.[161] Among the most important cultural performances held annually are the Lucerne Festival,[162] the Montreux Jazz Festival[163] and the Locarno International Film Festival.[164]

Alpine symbolism has played an essential role in shaping the history of the country and the Swiss national identity.[165] [166] Nowadays some concentrated mountain areas have a strong highly energetic ski resort culture in winter, and a hiking (wandering) or Mountain biking culture in summer. Other areas throughout the year have a recreational culture that caters to tourism, yet the quieter seasons are spring and autumn when there are fewer visitors. A traditional farmer and herder culture also predominates in many areas and small farms are omnipresent outside the cities. Folk art is kept alive in organisations all over the country. In Switzerland it is mostly expressed in music, dance, poetry, wood carving and embroidery. The alphorn, a trumpet-like musical instrument made of wood, has become alongside yodeling and the accordion an epitome of traditional Swiss music.[167] [168]

Literature

As the Confederation, from its foundation in 1291, was almost exclusively composed of German-speaking regions, the earliest forms of literature are in German. In the 18th century French became the fashionable language in Bern and elsewhere, while the influence of the French-speaking allies and subject lands was more marked than before.[170]

Jean-Jacques Rousseau was not only a writer but also an influential philosopher of the eighteenth-century[169] (his statue in Geneva).

Among the classics of Swiss German literature are Jeremias Gotthelf (1797–1854) and Gottfried Keller (1819–1890). The undisputed giants of 20th century Swiss literature are Max Frisch (1911–91) and Friedrich Dürrenmatt (1921–90), whose repertoire includes Die Physiker (The Physicists) and Das Versprechen (The Pledge), released in 2001 as a Hollywood film.[171]

Prominent French-speaking writers were Jean-Jacques Rousseau (1712–1778) and Germaine de Staël (1766–1817). More recent authors include Charles Ferdinand Ramuz (1878–1947), whose novels describe the lives of peasants and mountain dwellers, set in a harsh environment and Blaise Cendrars (born Frédéric Sauser, 1887–1961).[171] Also Italian and Romansh-speaking authors contributed but in more modest way given their small number.

Probably the most famous Swiss literary creation, *Heidi*, the story of an orphan girl who lives with her grandfather in the Alps, is one of the most popular children's books ever and has come to be a symbol of Switzerland. Her creator, Johanna Spyri (1827–1901), wrote a number of other books around similar themes.[171]

Media

The freedom of the press and the right to free expression is guaranteed in the federal constitution of Switzerland.[172] The Swiss News Agency (SNA) broadcasts information around-the-clock in three of the four national languages—on politics, economics, society and culture. The SNA supplies almost all Swiss media and a couple dozen foreign media services with its news.[172]

Switzerland has historically boasted the greatest number of newspaper titles published in proportion to its population and size.[173] The most influential newspapers are the German-language Tages-Anzeiger and Neue Zürcher Zeitung NZZ, and the French-language Le Temps, but almost every city has at least one local newspaper. The cultural diversity accounts for a large number of newspapers.[173]

In contrast to the print media, the broadcast media has always been under greater control of the government.[173] The Swiss Broadcasting Corporation, whose name was recently changed to SRG SSR idée suisse, is charged with the production and broadcast of radio and television programs. SRG SSR studios are distributed throughout the various language regions. Radio content is produced in six central and four regional studios while the television programs are produced in Geneva, Zurich and Lugano. An extensive cable network also allows most Swiss to access the programs from neighboring countries.[173]

Sports

Skiing, snowboarding and mountaineering are among the most popular sports in Switzerland, the nature of the country being particularly suited for such activities.[174] Winter sports are practiced by the natives and tourists since the second half of the 19th century with the invention of bobsleigh in St. Moritz.[175] The first world ski championships were held in Mürren (1931) and St. Moritz (1934). The latter town hosted the second Winter Olympic Games in 1928 and the fifth edition in 1948. Among the most successful skiers and world champions are Pirmin Zurbriggen and Didier Cuche.

Ski area over the glaciers of Saas-Fee

Spengler Cup in Davos

Many Swiss are fans of football and the national team or 'Nati' is widely supported. Switzerland was the joint host, with Austria, of the Euro 2008 tournament. Many Swiss also follow ice hockey and support one of the 12 clubs in the League A. In April 2009, Switzerland hosted the 2009 IIHF World Championship for the 10th time.[176] The National League A is the most attended league in Europe.[177] The numerous lakes make Switzerland an attractive place for sailing. The largest, Lake Geneva, is the home of the sailing team Alinghi which was the first European team to win the America's Cup in 2003 and which successfully defended the title in 2007. Tennis has become an increasingly popular sport, and Swiss players such as Martina Hingis and Roger Federer have won multiple Grand Slams.

Motorsport racecourses and events were banned in Switzerland following the 1955 Le Mans disaster with exception to events such as Hillclimbing. However, this ban was overturned in June 2007.[179] During this period, the country still produced successful racing drivers such as Clay Regazzoni, Sebastian Buemi, Jo Siffert and successful World Touring Car Championship driver Alain Menu. Switzerland also won the A1GP World Cup of Motorsport in 2007–08 with driver Neel Jani. Swiss motorcycle racer Thomas Lüthi won the 2005 MotoGP World Championship in the 125cc category.

In a eight-year span, Roger Federer has won a record 16 Grand Slam singles titles, making him the most successful men's tennis player ever.[178]

Traditional sports include Swiss wrestling or "Schwingen". It is an old tradition from the rural central cantons and considered the national sport by some. Hornussen is another indigenous Swiss sport, which is like a cross between baseball and golf.[180] Steinstossen is the Swiss variant of stone put, a competition in throwing a heavy stone. Practiced only among the alpine population since prehistoric times, it is recorded to have taken place in Basel in the 13th century. It is also central to the Unspunnenfest, first held in 1805, with its symbol the 83.5 kg stone named *Unspunnenstein*.[181]

Cuisine

The cuisine of Switzerland is multi-faceted. While some dishes such as fondue, raclette or rösti are omnipresent through the country, each region developed its own gastronomy according to the differences of climate and languages.[182] Traditional Swiss cuisine uses ingredients similar to those in other European countries, as well as unique dairy products and cheeses such as Gruyère or Emmental, produced in the valleys of Gruyères and Emmental. The number of fine-dining establishments is high, particularly in western Switzerland.[183] [184]

Chocolate had been made in Switzerland since the 18th century but it gained its reputation at the end of the 19th century with the invention of modern techniques such as conching and tempering which enabled its production on a high quality level. Also a breakthrough was the invention of milk chocolate in 1875 by Daniel Peter. The Swiss are the world's largest consumers of chocolate.[185] [186]

The most popular alcoholic drink in Switzerland is wine. Switzerland is notable for the variety of grapes grown because of the large variations in terroirs, with their specific mixes of soil, air, altitude and light. Swiss wine is produced mainly in Valais, Vaud (Lavaux), Geneva and Ticino, with a small majority of white wines. Vineyards have been cultivated in Switzerland since the Roman era, even though certain traces can be found of a more ancient origin. The most widespread varieties are the Chasselas (called Fendant in Valais) and Pinot Noir. The Merlot is the main variety produced in Ticino.[187] [188]

See also

- Outline of Switzerland
- Index of Switzerland-related articles

References

Notes

[1] *De jure* "federal city"; *de facto* capital. Because of historical federalist sensibilities, Swiss law does not designate a formal capital, and some federal institutions such as courts are located in other cities.

[2] Federal Constitution, article 4 (http://www.admin.ch/ch/e/rs/101/a4.html), "National languages" : *National languages* are German, French, Italian and Romansh; Federal Constitution, article 70 (http://www.admin.ch/ch/e/rs/101/a70.html), "Languages", paragraph 1: The *official languages* of the Confederation are German, French and Italian. Romansh shall be an official language for communicating with persons of Romansh language.

[3] Traditional. The Federal Charter mentions only "early August" and the treaty is a renewal of an older one, now lost.

[4] A solemn declaration of the Tagsatzung (http://www.verfassungen.de/ch/tagsatzungsbeschluss48.htm) declared the Federal Constitution adopted on 12 September 1848. A resolution of the Tagsatzung (http://www.verfassungen.de/ch/tagsatzungsbeschluss48-2.htm) of 14 September 1848 specified that the powers of the institutions provided for by the 1815 Federal Treaty would expire at the time of the constitution of the Federal Council, which took place on 16 November 1848.

[5] "Press Release: Provisional findings on population growth in Switzerland in 2011 Demographic growth continues" (http://www.bfs.admin.ch/bfs/portal/en/index/themen/01/22/press.html?pressID=7951). *Swiss Federal Statistical Office*. Swiss Federal Statistical Office, Neuchâtel. 2011. . Retrieved 2012-04-26.

[6] "Switzerland" (http://www.imf.org/external/pubs/ft/weo/2012/01/weodata/weorept.aspx?pr.x=57&pr.y=9&sy=2009&ey=2012&scsm=1&ssd=1&sort=country&ds=.&br=1&c=146&s=NGDPD,NGDPDPC,PPPGDP,PPPPC,LP&grp=0&a=). International Monetary Fund. . Retrieved 2012-04-22.

[7] "Human Development Report 2011" (http://hdr.undp.org/en/media/HDR_2010_EN_Complete.pdf). United Nations. 2011. . Retrieved 4 November 2010.

[8] The Swiss German name is also sometimes spelled as *Schwyz* or *Schwiiz*. *Schwyz* is also the standard German (and international) name of one of the Swiss cantons.

[9] In Western Europe or Central Europe, depending on the definition. See Geography of Switzerland.

[10] Thomas Fleiner, Alexander Misic, Nicole Töpperwien, *Swiss Constitutional Law*, p. 28, Kluwer Law International

[11] "US is still by far the richest country, China fastest growing" (http://www.digitaljournal.com/article/298716). Digitaljournal.com. . Retrieved 2010-11-08.

[12] Franc's rise puts Swiss top of rich list (http://www.guardian.co.uk/business/2011/oct/19/currency-appreciation-makes-swiss-wealthiest) Simon Bowers, guardian.co.uk, Wednesday 19 October 2011

[13] "Swiss and German cities dominate ranking of best cities in the world" (http://www.citymayors.com/features/quality_survey.html). Citymayors.com. 2009-04-28. . Retrieved 2010-04-29.
[14] Constitutional Patriotism and Exclusion: the Swiss Case (http://www.euroculturemaster.org/pdf/groep1_2/Mangiat.pdf) euroculturemaster.org. Retrieved on 2009-07-30
[15] In Search of Natural Identity: Alpine Landscape and the Reconstruction of the Swiss Nation. (http://journals.cambridge.org/download.php?file=/CSS/CSS40_04/S0010417598001686a.pdf&code=58775d5e7c8f256fef0a6e06d5ccf9fa) journals.cambridge.org. Retrieved on 2009-07-30
[16] OED Online Etymology Dictionary (http://www.etymonline.com/index.php?term=Swiss) etymonline.com. Retrieved on 2009-06-25
[17] Room, Adrian. *Placenames of the World*. London: MacFarland and Co., Inc., 1997.
[18] Switzerland, the Catholic Encyclopedia (http://www.newadvent.org/cathen/14358a.htm) newadvent.org. Retrieved on 2010-01-26
[19] On Schwyzers, Swiss and Helvetians, Federal Department of Home Affairs, bar.admin.ch (http://www.bar.admin.ch/archivgut/00591/00601/00602/index.html?lang=en#). Retrieved on 2010-01-26
[20] Züritütsch, Schweizerdeutsch (p. 2) (http://www.schweizer-deutsch.ch/uploads/media/einige_Beispielseiten_2007-07-05_01.pdf) schweizerdeutsch.ch. Retrieved on 20010-01-26
[21] Kanton Schwyz: Kurzer historischer Überblick (http://www.sz.ch/xml_1/internet/de/application/d2/d56/d756/f759.cfm) sz.ch. Retrieved on 2010-01-26
[22] Marco Marcacci, Confederatio helvetica (2002) (http://hls-dhs-dss.ch/textes/d/D9827.php), Historical Lexicon of Switzerland.
[23] *Helvetia* in German (http://www.hls-dhs-dss.ch/textes/d/D16440.php), French (http://www.hls-dhs-dss.ch/textes/f/F16440.php) and Italian (http://www.hls-dhs-dss.ch/textes/i/I16440.php) in the online *Historical Dictionary of Switzerland*.
[24] History (http://www.swissworld.org/en/history/prehistory_to_romans/prehistoric_times/) swissworld.org. Retrieved on 2009-06-27
[25] Switzerland's Roman heritage comes to life (http://www.swissinfo.ch/eng/swissinfo.html?siteSect=108&sid=6054935&cKey=1126454871000) swissinfo.ch
[26] Switzerland history (http://www.nationsencyclopedia.com/Europe/Switzerland-HISTORY.html) Nationsencyclopedia.com. Retrieved on 2009-11-27
[27] History of Switzerland (http://www.nationsonline.org/oneworld/History/Switzerland-history.htm) Nationsonline.org. Retrieved on 2009-11-27
[28] Schwabe & Co.: *Geschichte der Schweiz und der Schweizer*, Schwabe & Co 1986/2004. ISBN 3-7965-2067-7 (German)
[29] A Brief Survey of Swiss History (http://www.eda.admin.ch/eda/en/home/reps/ocea/vaus/infoch/chhist.html) admin.ch, Retrieved on 2009-06-22
[30] *Swiss border* in German (http://www.hls-dhs-dss.ch/textes/d/D7841.php), French (http://www.hls-dhs-dss.ch/textes/f/F7841.php) and Italian (http://www.hls-dhs-dss.ch/textes/i/I7841.php) in the online *Historical Dictionary of Switzerland*.
[31] *Bundesstadt* in German (http://www.hls-dhs-dss.ch/textes/d/D10102.php), French (http://www.hls-dhs-dss.ch/textes/f/F10102.php) and Italian (http://www.hls-dhs-dss.ch/textes/i/I10102.php) in the online *Historical Dictionary of Switzerland*.
[32] Tunnel Vision: Switzerland's AlpTransit Gotthard Tunnel (http://www.inboundlogistics.com/articles/global/global0106a.shtml) inboundlogistics.com. Retrieved on 2010-04-24
[33] *Histoire de la Suisse*, Éditions Fragnière, Fribourg, Switzerland
[34] Lenin and the Swiss non-revolution (http://www.swissinfo.ch/eng/specials/extraordinary_exiles/index/Lenin_and_the_Swiss_non-revolution.html?cid=12812) swissinfo.ch. Retrieved on 2010-01-25
[35] *Let's Swallow Switzerland* by Klaus Urner (Lexington Books, 2002).
[36] Book review: Target Switzerland: Swiss Armed Neutrality in World War II, Halbrook, Stephen P. (http://stonebooks.com/archives/981111.shtml) stonebooks.com. Retrieved on 2009-12-02
[37] *Asylum* in German (http://www.hls-dhs-dss.ch/textes/d/D10374.php), French (http://www.hls-dhs-dss.ch/textes/f/F10374.php) and Italian (http://www.hls-dhs-dss.ch/textes/i/I10374.php) in the online *Historical Dictionary of Switzerland*.
[38] The Bergier Commission Final Report (http://www.uek.ch/en/schlussbericht/synthesis/ueke.pdf), page 498.
[39] The Bergier Commission Final Report (http://www.uek.ch/en/schlussbericht/synthesis/ueke.pdf), page 521.
[40] Country profile: Switzerland (http://www.fco.gov.uk/en/travel-and-living-abroad/travel-advice-by-country/country-profile/europe/switzerland?profile=today) UK Foreign and Commonwealth Office, Retrieved on 2009-11-25
[41] Geography (http://www.swissworld.org/en/geography/swiss_geography/contrasts/) swissworld.org, Retrieved on 2009-06-23
[42] "Landscape and Living Space" (http://www.eda.admin.ch/eda/en/home/reps/ocea/vaus/infoch/chgeog.html). *Federal Department of Foreign Affairs*. Federal Administration admin.ch. 2007-07-31. . Retrieved 2009-06-25.
[43] A zoomable map of Switzerland is available at either swissinfo-geo.org (http://www.swissinfo-geo.org/) or swissgeo.ch (http://www.swissgeo.ch/); a zoomable satellite picture is at map.search.ch (http://map.search.ch/).
[44] Physical Geography of Switzerland (http://www.bfs.admin.ch/bfs/portal/en/index/regionen/thematische_karten/maps/raum_und_umwelt/naturraum_schweiz.html) bfs.admin.ch. Retrieved on 2010-05-06
[45] Ice volume of Switzerland's glaciers calculated (http://www.sciencecentric.com/news/article.php?q=09062217-ice-volume-switzerland-glaciers-calculated) sciencecentric.com. Retrieved on 2010-05-06
[46] Landscape and climate (http://www.cp-pc.ca/english/switzerland/) cp-pc.ca. Retrieved on 2009-12-14
[47] Herbermann, Charles George (1913). *The Catholic Encyclopedia*. Encyclopedia Press. p. 358.
[48] Swiss Alps Jungfrau-Aletsch (http://whc.unesco.org/en/list/1037/) unesco.org

[49] Climate in Switzerland (http://www.about.ch/geography/climate/index.html) about.ch, Retrieved on 2009-06-23
[50] Country guide, Switzerland (http://www.bbc.co.uk/weather/world/country_guides/results.shtml?tt=TT005790) bbc.co.uk, Retrieved on 2009-11-20
[51] "Climate data sunshine/temperature/rainfall choosable on federal weather bureau" (http://www.meteoschweiz.admin.ch/web/en/climate/climate_today/swiss_climate_maps.html). Meteoschweiz.admin.ch. 2011-05-23. . Retrieved 2011-05-29.
[52] The strengths of Switzerland and its people, p.3 (http://www.image-switzerland.ch/fileadmin/user_upload/pdf/e/Image_der_Schweiz/Switzerland_s_strengths_people.pdf) Federal Department of Foreign Affairs FDFA. Retrieved on 2009-12-02
[53] Impact of climate change (http://www.swissworld.org/en/environment/climate_change/impact_of_climate_change/) swissworld.org. Retrieved on 2010-01-14.
[54] As shown in this image, the current members of the council are (as of January 2012, from left to right): Federal Councillor Johann Schneider-Ammann, Federal Councillor Didier Burkhalter, Vice-President Ueli Maurer, President Eveline Widmer-Schlumpf, Federal Councillor Doris Leuthard, Federal Councillor Simonetta Sommaruga, Federal Councillor Alain Berset and Federal Chancellor Corina Casanova
[55] Political System (http://www.eda.admin.ch/eda/en/home/reps/ocea/vaus/infoch/chpoli.html) admin.ch, Retrieved on 2009-06-22
[56] "Political System" (http://www.eda.admin.ch/eda/en/home/reps/eur/vgbr/infoch/chpoli.html). Federal Department of Foreign Affairs. .
[57] The SVP/UDC has suffered a split since the election, with both their councillors defecting to the Conservative Democratic Party of Switzerland (BDP/PBD). As of 2009, with the election of Ueli Maurer, the SVP/UDC and the BDP/PBD hold one seat each.
[58] The Judiciary: The Federal Supreme Court (http://www.ch.ch/behoerden/00215/00329/00353/index.html?lang=en) ch.ch. Retrieved on 2009-12-15
[59] Cantons (http://www.ch.ch/behoerden/00215/00330/index.html?lang=en) ch.ch. Retrieved on 2009-12-15
[60] How direct democracy makes Switzerland a better place (http://www.telegraph.co.uk/news/1435383/How-direct-democracy-makes-Switzerland-a-better-place.html) telegraph.co.uk. Retrieved on 2009-12-04
[61] Since 1999, an initiative can also be in the form of a general proposal to be elaborated by Parliament, but because it is considered less attractive for various reasons, this form of initiative has yet to find any use.
[62] That is a majority of 23 cantonal votes, because the result of the popular vote in the six traditional half-cantons each counts as half the vote of one of the other cantons.
[63] Tremblay; Lecours; et al. (2004) Mapping the Political Landscape. Toronto: Nelson.
[64] Turner; Barry (2001). The Statement's Yearbook. New York: MacMillan Press ltd.
[65] Banks, Arthur (2006). Political Handbook of The World 2005–2006. Washington: Cq Press.
[66] Enclaves of the world (http://enclaves.webs.com/westerneurope.htm) enclaves.webs.com. Retrieved on 2009-12-15
[67] Vorarlberg (http://www.austria.org/content/view/74/) austria.org. Retrieved on 2009-12-15
[68] c2d@zda.uzh.ch (2008-08-20). "unige.ch – Direct democracy in the world" (http://www.c2d.ch/?entit=10&vote=101&lang=). C2d.ch. . Retrieved 2010-04-29.
[69] Neutrality and isolationism (http://www.swissworld.org/en/politics/foreign_policy/neutrality_and_isolationism/) swissworld.org, Retrieved on 2009-06-23
[70] "Switzerland – Country history and economic development" (http://www.nationsencyclopedia.com/economies/Europe/Switzerland-COUNTRY-HISTORY-AND-ECONOMIC-DEVELOPMENT.html). *nationsencyclopedia.com*. . Retrieved 12 December 2009.
[71] Henri Dunant, the Nobel Peace Prize 1901 (http://nobelprize.org/nobel_prizes/peace/laureates/1901/dunant-bio.html) nobelprize.org. Retrieved on 2009-12-02
[72] Sports directory (http://www.ifsports-guide.ch/english/navigation/annuaire_en.html) if-sportsguide.ch. Retrieved on 2010-01-25
[73] An initiative to abandon this practice (http://www.schutz-vor-waffengewalt.ch/) has been launched on 4 September 2007, and supported by GSoA, the Green Party of Switzerland and the Social Democratic Party of Switzerland as well as other organisations which are listed at schutz-vor-waffengewalt.ch (http://www.schutz-vor-waffengewalt.ch/organisationen.html)
[74] "Zwei Drittel der Rekruten diensttauglich (Schweiz, NZZ Online)" (http://www.nzz.ch/nachrichten/schweiz/zwei_drittel_der_rekruten_diensttauglich_1.687233.html). . Retrieved 23 February 2009.
[75] Armeezahlen www.vbs.admin.ch (http://www.vbs.admin.ch/internet/vbs/de/home/documentation/armeezahlen/eff.html) (German)
[76] As context, according to Edwin Reischauer, "To be neutral you must be ready to be highly militarized, like Switzerland or Sweden." – *see* Chapin, Emerson. "Edwin Reischauer, Diplomat and Scholar, Dies at 79," (http://query.nytimes.com/gst/fullpage.html?res=9C0CE2DC163AF931A3575AC0A966958260&scp=1&sq=edwin+o+reischauer&st=nyt) *New York Times*. September 2, 1990.
[77] Volksabstimmung vom 26. November 1989 (http://www.admin.ch/ch/d//pore/va//19891126/index.html) admin.ch. Retrieved on 2010-01-25
[78] *L'évolution de la politique de sécurité de la Suisse* ("Evolution of Swiss Security Policies") by Manfred Rôsch, NATO.int (http://www.nato.int/docu/revue/1993/9306-05.htm)
[79] Volksinitiative 'für eine glaubwürdige Sicherheitspolitik und eine Schweiz ohne Armee (in German) (http://www.admin.ch/ch/d//pore/va/20011202/det482.html) admin.ch. Retrieved on 2009-12-07
[80] "Watches" (http://www.swissworld.org/en/economy/key_sectors/watches/). Swissworld.org. . Retrieved 2012-02-28.
[81] https://www.credit-suisse.com/news/en/media_release.jsp?ns=41874
[82] https://www.credit-suisse.com/upload/news-live/000000022454.pdf

[83] 2010 Index of Economic Freedom: Switzerland (http://www.heritage.org/index/country/Switzerland) heritage.org. Retrieved on 2011-01-25
[84] https://www.cia.gov/library/publications/the-world-factbook/rankorder/2004rank.html
[85] "World Economic Forum – Global Competitiveness Report" (http://www.weforum.org/en/initiatives/gcp/Global Competitiveness Report/index.htm). Weforum.org. . Retrieved 2010-04-29.
[86] http://ec.europa.eu/research/innovation-union/pdf/iu-scoreboard-2010_en.pdf
[87] Taylor & Francis Group (2002). *Western Europe*. Routledge. pp. 645–646. ISBN 1-85743-152-9.
[88] The most powerful cities in the world (http://www.citymayors.com/economics/power-cities.html) citymayors.com. Retrieved on 2012-04-27
[89] "Six Swiss companies make European Top 100" (http://www.swissinfo.ch/eng/business/detail/Six_Swiss_companies_make_European_Top_100.html?siteSect=161&sid=7174196&cKey=1161172317000). swissinfo.ch. 18 October 2008. . Retrieved 22 July 2008.
[90] Swiss Statistical Yearbook 2008 by Swiss Federal Statistical Office
[91] Swiss jobless reach 12-year high – a mere 4.4 pct (http://www.businessweek.com/ap/financialnews/D9D3FGVO0.htm)
[92] Policy Brief: Economic Survey of Switzerland, 2007 (http://www.oecd.org/dataoecd/39/8/39539300.pdf) (326 KiB), OECD
[93] Economic Policy Reforms: Going for Growth 2008 – Switzerland Country Note (http://www.oecd.org/dataoecd/29/49/40202407.pdf) (45 KiB)
[94] Domestic purchasing power of wages (http://www.osec.ch/internet/osec/en/home/invest/factors/infrastructure/live/costs.-RelatedBoxSlot-47301-ItemList-89920-File.File.pdf/Domestic purchasing power of wages E.pdf) (68 KiB)
[95] Switzerland tops in buying power (http://goliath.ecnext.com/coms2/gi_0199-4179358/Switzerland-tops-in-buying-power.html) goliath.ecnext.com. Retrieved on 2010-01-14.
[96] Want the world's best wages? Move to Switzerland (http://www.reuters.com/article/idUSTRE57Q23H20090827) reuters.com. Retrieved on 2010-01-14.
[97] The Swiss education system (http://www.swissworld.org/en/education/general_overview/the_swiss_education_system/) swissworld.org, Retrieved on 2009-06-23
[98] ETH Zurich (Swiss Federal Institute of Technology) (http://www.topuniversities.com/university/199/eth-zurich-swiss-federal-institute-of-technology) topouniversities.com. Retrieved on 2010-04-30
[99] Top.Universities (http://www.topuniversities.com/world-university-rankings) Retrieved on 2010-04-30
[100] In 2008, the ETH Zurich was ranked 15th in the field *Natural Sciences and Mathematics* by the Shanghai Academic Ranking of World Universities and the EPFL in Lausanne was ranked 18th in the field *Engineering/Technology and Computer Sciences* by the same ranking.
[101] "Shanghai Ranking 2008 Top 100 world universities in Natural Sciences and Mathematics" (http://ed.sjtu.edu.cn/ARWU-FIELD2008/SCI2008.htm). Ed.sjtu.edu.cn. . Retrieved 2010-11-02.
[102] Education at Glance 2005 (http://www.ecs.org/html/offsite.asp?document=http://www.oecd.org/dataoecd/20/25/35345692.pdf) by the OECD: Percentage of foreign students in tertiary education.
[103] (http://www.studyihub.com/the-graduate-institute-of-international-studies-geneva/)
[104] (http://www.law.umn.edu/eperspectives/summer2010/news-and-events/kendra-magraw-10-accepted-at-genevas-prestigious-iheid/)
[105] John Snygg, *A New Approach to Differential Geometry Using Clifford's Geometric Algebra* (http://books.google.ch/books?id=qlX3CqMUHpUC&printsec=frontcover#v=onepage&q&f=false), *p. 38*
[106] Dr. phil. Roland Mueller. "Muellerscience.com" (http://www.muellerscience.com/SPEZIALITAETEN/Schweiz/SchweizerNobelpreistraeger.htm). Muellerscience.com. . Retrieved 2011-05-29.
[107] Nobel prizes in non-science categories included.
[108] "Mueller Science – Spezialitaeten: Schweizer Nobelpreisträger" (http://www.muellerscience.com/SPEZIALITAETEN/Schweiz/SchweizerNobelpreistraeger.htm). . Retrieved 31 July 2008.
[109] info.cern.ch (http://info.cern.ch/) Retrieved on 2010-04-30
[110] "CERN - the largest laboratory in the world www.swissworld.org" (http://www.swissworld.org/en/switzerland/resources/story_switzerland/cern_the_largest_laboratory_in_the_world/). Swissworld.org. . Retrieved 2010-04-29.
[111] "How to get to CERN" (http://user.web.cern.ch/user/Institute/Integration/GetToCERN/GetToCERN.html). .
[112] Company overview www.oerlikon.com (http://www.oerlikon.com/ecomaXL/index.php?site=SPACE_EN_company_overview)
[113] maxon motor ag (2004-01-04). "Media releases" (http://www.maxonmotor.ch/ch/en/media_releases_5619.html). Maxonmotor.ch. . Retrieved 2010-04-29.
[114] Prof Clive Church (may 2003). "The contexts of Swiss opposition to Europe" (http://www.webcitation.org/5lZameh86) (PDF, 124 KiB). Sussex European Institute. p. 12. Archived from the original (http://www.sussex.ac.uk/sei/documents/wp64.pdf) on 2009-11-26. . Retrieved 13 June 2008.
[115] "Volksinitiative "Ja zu Europa!" *(Initiative "Yes to Europe!")"* (http://www.bfs.admin.ch/bfs/portal/de/index/themen/17/03/blank/key/2001/01.Document.22675.pdf) (in German) (PDF, 1.1 MiB). BFS/OFS/UST. 13 February 2003. . Retrieved 15 June 2008.
[116] "Volksinitiative "Ja zu Europa!", nach Kantonen. (Initiative "Yes to Europe!" by Canton)." (http://www.bfs.admin.ch/bfs/portal/de/index/themen/17/03/blank/key/2001/01.Document.85488.xls) (in German) (XLS). BFS/OFS/UST. 16 January 2003. . Retrieved 15 June 2008.

[117] Switzerland and the European Union (http://www.europa.admin.ch/themen/00499/00503/index.html?lang=en) europa.admin.ch. Retrieved on 2010-01-25

[118] Switzerland in Schengen: end to passport checks (http://www.euronews.net/2008/12/11/switzerland-in-schengen-end-to-passport-checks/) euronews.net. Retrieved on 2010-01-25

[119] "Vote No. 502 – Summary" (http://www.admin.ch/ch/d/pore/va/20030518/det502.html) (in German). 18 May 2003. .

[120] "Vote No. 501 – Summary" (http://www.admin.ch/ch/d/pore/va/20030518/det501.html) (in German). 18 May 2003. .

[121] "Federal government energy research" (http://www.bfe.admin.ch/forschungnetze/01223/index.html?lang=en). 16 January 2008. .

[122] "Switzerland to Phase Out Nuclear Energy; E.U. Strikes Deal on 'Stress Tests'" (http://news.sciencemag.org/scienceinsider/2011/05/switzerland-to-phase-out-nuclear.html). 25 May 2011. .

[123] "Verkehrsleistungen – Daten, Indikatoren admin.ch (German)" (http://www.bfs.admin.ch/bfs/portal/de/index/themen/11/05/blank/key/verkehrsleistungen/mengen.html). Bfs.admin.ch. 2010-03-25. . Retrieved 2010-04-29.

[124] Schienenverkehr (http://www.bav.admin.ch/dokumentation/publikationen/00475/01623/01624/index.html?lang=de) admin.ch (German)

[125] Rhaetian Railway in the Albula / Bernina Landscapes (http://whc.unesco.org/en/list/1276/) unesco.org

[126] Swiss sit atop ranking of greenest nations (http://www.msnbc.msn.com/id/22802081/) msnbc.com. Retrieved on 2009-12-02

[127] Party grouping (http://unfccc.int/parties_and_observers/parties/negotiating_groups/items/2714.php) unfccc.int. Retrieved on 2009-12-02

[128] W3design. "Swiss Recycling" (http://www.swissrecycling.ch/deutsch/wregel.htm). Swissrecycling.ch. . Retrieved 2010-04-29.

[129] Stadtreinigung Basel-Stadt (http://www.stadtreinigung-bs.ch/page.php?lang=0&sel=114)—Pricelist bags and stickers

[130] "Recycling around the world" (http://news.bbc.co.uk/1/hi/world/europe/4620041.stm). BBC. 25 June 2005. . Retrieved 24 April 2006.

[131] Richtig Entsorgen (Kanton Basel-Stadt) (http://www.stadtreinigung-bs.ch/data/0d1b64Sauberbuch2004.pdf) (1.6 MiB)—Wilde Deponien sind verboten... Für die Beseitigung widerrechtlich deponierter Abfälle wird zudem eine Umtriebsgebühr von Fr. 200.– oder eine Busse erhoben (page 90)

[132] History of paper manufacturing (http://www.geschichteinchronologie.ch/welt/papiergeschichte.html) in German, retrieved 03-05-2011

[133] Swiss Federal Statistical Office. "Languages and religions – Data, indicators" (http://www.bfs.admin.ch/bfs/portal/en/index/themen/01/05/blank/key/sprachen.html). . Retrieved 9 October 2007. The first number refers to the share of languages within total population. The second refers to the Swiss citizens only.

[134] The Parliamentary Services (http://www.parlament.ch/e/service-presse/parlamentsdienste/aufgaben/Pages/default.aspx) parlament.ch. Retrieved on 2010-01-25

[135] "SBB: Billette – OnlineTicket" (http://mct.sbb.ch/mct/reisemarkt/billette/online-ticket.htm). Mct.sbb.ch. . Retrieved 2010-04-29.

[136] Minorities and bilingualism (http://www.swissworld.org/en/people/language/minorities_and_bilingualism/) swissworld.org. Retrieved on 2010-01-25

[137] Ausländerinnen und Ausländer in der Schweiz – Bericht 2008 (German) (http://www.bfs.admin.ch/bfs/portal/de/index/themen/01/22/publ.Document.114724.pdf) (1196 KiB), Swiss Federal Statistical Office, page 12.

[138] Ausländerinnen und Ausländer in der Schweiz – Bericht 2008 (German) (http://www.bfs.admin.ch/bfs/portal/de/index/themen/01/22/publ.Document.114724.pdf) (1196 KiB), Swiss Federal Statistical Office, page 72.

[139] Foreign population in Switzerland detailed by nationality, 1980–2006 (German) (http://www.bfs.admin.ch/bfs/portal/de/index/themen/01/07/blank/key/01/01.Document.67321.xls), Swiss Federal Statistical Office.

[140] Definitive report on racism in Switzerland by UN expert (http://www.humanrights.ch/en/Switzerland/Human-Rights-in-Internal-Affairs/Racism/Studies/idart_5119-content.html) humanrights.ch

[141] Switzerland (http://www.who.int/countries/che/en/index.html) who.int. Retrieved on 2009-06-29

[142] Life expectancy at birth, 2006 (http://apps.who.int/whosis/database/country/compare.cfm?strISO3_select=CHE&strIndicator_select=LEX0Male,LEX0Female&language=english&order_by=FirstValue DESC) who.int. Retrieved on 2009-06-29

[143] OECD Health Data 2006 (http://www.oecd.org/dataoecd/29/52/36960035.pdf) oecd.org. Retrieved on 2009-06-29

[144] OECD and WHO survey of Switzerland's health system (http://www.oecd.org/document/47/0,2340,en_2649_201185_37562223_1_1_1_1,00.html) oecd.org. Retrieved on 2009-06-29

[145] Where people live (http://www.swissworld.org/en/geography/town_and_country_planning/where_people_live/) swissworld.org. Retrieved on 2009-06-26

[146] Städte und Agglomerationen unter der Lupe (http://www.are.admin.ch/dokumentation/00121/00224/index.html?lang=de&msg-id=27412) admin.ch. Retrieved on 2009-06-26

[147] Swiss countryside succumbs to urban sprawl (http://www.swissinfo.ch/eng/front/Swiss_countryside_succumbs_to_urban_sprawl.html?siteSect=106&sid=9823369&cKey=1223485367000&ty=st) swissinfo.ch. Retrieved on 2009-06-30

[148] Enquête représentative sur l'urbanisation de la Suisse (Pronatura) (http://www.gfs-zh.ch/content.php?pid=201) gfs-zh.ch. Retrieved on 2009-06-30

[149] Swiss plateau (http://www.swissworld.org/en/geography/the_three_regions/the_swiss_plateau/) swissworld.org. Retrieved on 2009-06-29

[150] Quality of living (http://www.mercer.com/qualityofliving) mercer.com. Retrieved on 2009-06-26

[151] International Religious Freedom Report 2004 – Switzerland (http://www.state.gov/g/drl/rls/irf/2004/35487.htm), U.S. Department of State.

[152] "CIA World Factbook section on Switzerland" (https://www.cia.gov/library/publications/the-world-factbook/geos/sz.html#People). Cia.gov. . Retrieved 2011-05-29.

[153] Bovay, Claude; Raphaël Broquet (December 2004). "Introduction" (http://www.bfs.admin.ch/bfs/portal/fr/index/news/publikationen.Document.50517.pdf) (in French) (PDF). *Recensement fédéral de la population 2000*. Neuchâtel: Federal Statistical Office. p. 132. ISBN 3-303-16074-0. . Retrieved 21 August 2010.

[154] Social values, Science and Technology (http://ec.europa.eu/public_opinion/archives/ebs/ebs_225_report_en.pdf)PDF (1.64 MiB), Eurobarometer, June 2005.

[155] Greeley, Andrew. 2003. Religion in Europe at the End of the Second Millennium. New Brunswick, New Jersey: Transaction Publishers

[156] Reclus, Élisée (1881). *The Earth and Its Inhabitants*. D. Appleton and Company. p. 478.

[157] Volksabstimmung vom 2. März 1980 (http://www.admin.ch/ch/d/pore/va/19800302/index.html) admin.ch. Retrieved on 2010

[158] Swiss culture (http://www.swissworld.org/en/culture/) swissworld.org. Retrieved on 2009-12-01

[159] European Year of Intercultural Dialogue (http://delche.ec.europa.eu/fileadmin/user_upload/Homepage_Dokumente/Presse_und_Information/Reden/2008-11-13_European_Year_of_Intercultural_Dialogue_2008__AFS_Intercultural_Programmes_Kaefigturm.pdf) Dr Michael Reiterer. Retrieved on 2009-12-01

[160] Switzerland: culture (http://www.traveldocs.com/ch/culture.htm) traveldocs.com. Retrieved on 2009-12-01

[161] Museums (http://www.swissworld.org/en/leisure/relaxation/museums/) swissworld.org. Retrieved on 2009-12-02

[162] Lucerne Festival (http://topics.nytimes.com/top/reference/timestopics/subjects/l/lucerne_festival/index.html) nytimes.com. Retrieved on 2010-12-15

[163] Switzerland's Prestigious Music Festival-Montreux Jazz Festival (http://www.nowpublic.com/culture/switzerland-s-prestigious-music-festival-montreux-jazz-festival) nowpublic.com. Retrieved on 2009-12-02

[164] Film festivals (http://www.swissworld.org/en/culture/film/film_festivals/) swissworld.org. Retrieved on 2009-12-02

[165] Mountains and hedgehogs (http://www.swissworld.org/en/culture/swissness/mountains_and_hedgehogs/) swissworld.org. Retrieved on 2009-12-01

[166] In Search of Natural Identity: Alpine Landscape and the Reconstruction of the Swiss Nation (http://journals.cambridge.org/download.php?file=/CSS/CSS40_04/S0010417598001686a.pdf&code=f5973a22cda752eed69390f92ddf4fcd), Oliver Zimmer, London School of Economics and Political Science

[167] Folk music (http://www.swissworld.org/en/culture/music/folk_music/) swissworld.org. Retrieved on 2009-12-02

[168] Culture of Switzerland (http://www.europe-cities.com/en/648/switzerland/culture/) europe-cities.com. Retrieved on 2009-12-14

[169] Art in literature (http://www.cp-pc.ca/english/switzerland/) cp-pc.ca. Retrieved on 2009-12-14

[170] From Encyclopædia Britannica Eleventh Edition, Swiss literature

[171] Literature (http://www.swissworld.org/en/culture/literature/german_speaking_authors/) swissworld.org, Retrieved on 2009-06-23

[172] Press and the media (http://www.ch.ch/private/00085/00090/00479/00480/index.html?lang=en) ch.ch. Retrieved on 2009-06-25

[173] Press in Switzerland (http://www.pressreference.com/Sw-Ur/Switzerland.html) pressreference.com. Retrieved on 2009-06-25

[174] Sport in Switzerland (http://www.europe-cities.com/en/672/switzerland/sport/) europe-cities.com. Retrieved on 2009-12-14

[175] A brief history of bobsleigh (http://www.fibt.com/index.php?id=39) fibt.com. Retrieved on 2009-11-02

[176] "IIHF World Championships 2009 official website" (http://www.iihf.com/channels/iihf-world-championship-oc09/home-oc/tournament-information.html). Iihf.com. 2009-05-10. . Retrieved 2010-04-29.

[177] Geering. "Hockeyarenas.net" (http://www.hockeyarenas.net/index.php3?ctID=ch&size=0&page=0701&c=eu&ctID=eu&atType=0&show=25&tdSaison=2011). Hockeyarenas.net. . Retrieved 2011-11-03.

[178] Roger Federer's Grand Slam Titles (http://sportsillustrated.cnn.com/vault/gallery/featured/GAL1157546/1/index.htm) sportsillustrated.cnn.com. Retrieved on 2010-06-14

[179] "Switzerland lifts ban on motor racing" (http://en.wikinews.org/wiki/Switzerland_lifts_ban_on_motor_racing). GrandPrix.com & DueMotori.com. 6 June 2007. . Retrieved 23 September 2008.

[180] Hornussen (http://www.swissroots.org/swissroots/en/stories/heritage/Heritage/Swiss Customs/Hornussen Baseball Swissstyle) swissroots.org. Retrieved on 2010-01-25

[181] Tradition and history (http://www.interlaken.ch/erlebnisse/sommer/festivals/unspunnenfest/tradition-und-geschichte.html?L=3) interlaken.ch. Retrieved on 2010-01-25

[182] Flavors of Switzerland (http://www.theworldwidegourmet.com/countries/flavors-of-switzerland/) theworldwidegourmet.com. Retrieved on 2009-06-24

[183] Michelin Guide Switzerland 2010 attests to the high quality of gourmet cooking with one new 2 star restaurant and 8 new one star (http://servicesv2.webmichelin.com/frontnews/servlet/GetElement?elementCode=56670) Press information, Michelin. Retrieved on 2009-12-14

[184] Swiss region serves up food with star power (http://www.usatoday.com/travel/destinations/2005-04-26-swiss-food_x.htm) usatoday.com. Retrieved on 2009-12-14

[185] Chocolate (http://www.swissworld.org/en/switzerland/swiss_specials/swiss_chocolate/swiss_breakthroughs/) swissworld.org. Retrieved on 2009-06-24

[186] Swiss Chocolate (http://www.germanworldonline.com/index.php/swisschocolatehistory/) germanworldonline.com. Retrieved on 2010-06-14

[187] Wine-producing Switzerland in short (http://www.swisswine.ch/english/bienv/main.asp) swisswine.ch. Retrieved on 2009-06-24

[188] Table 38. Top wine consuming nations per capita, 2006 (http://www.winebiz.com.au/statistics/world.asp) winebiz.com. Retrieved on 2010-06-14

References

Bibliography

- Church, Clive H. (2004) *The Politics and Government of Switzerland*. Palgrave Macmillan. ISBN 0-333-69277-2.
- Dalton, O.M. (1927) *The History of the Franks, by Gregory of Tours*. Oxford: The Clarendon Press.
- Fahrni, Dieter. (2003) *An Outline History of Switzerland. From the Origins to the Present Day*. 8th enlarged edition. Pro Helvetia, Zürich. ISBN 3-908102-61-8
- Historical Dictionary of Switzerland (2002–). Published electronically and in print simultaneously in three national languages of Switzerland.

External links

Government

- The Federal Authorities of the Swiss Confederation (http://www.admin.ch/)
- Chief of State and Cabinet Members (https://www.cia.gov/library/publications/world-leaders-1/world-leaders-s/switzerland.html)
- Swissworld.org, Switzerland's information portal (http://www.swissworld.org/)
- Swiss Statistics (http://www.bfs.admin.ch/bfs/portal/en/index.html), official website of the Swiss Federal Statistical Office

Reference

- Switzerland (https://www.cia.gov/library/publications/the-world-factbook/geos/sz.html) entry at *The World Factbook*
- Switzerland (http://www.britannica.com/EBchecked/topic/577225/Switzerland) entry at *Encyclopædia Britannica*
- Switzerland (http://ucblibraries.colorado.edu/govpubs/for/switzerland.htm) at *UCB Libraries GovPubs*
- Switzerland (http://www.dmoz.org/Regional/Europe/Switzerland/) at the Open Directory Project
- Key Development Forecasts for Switzerland (http://www.ifs.du.edu/ifs/frm_CountryProfile.aspx?Country=CH) from International Futures

Geography

- Wikimedia Atlas of Switzerland
- Switzerland travel guide from Wikitravel
- Federal Office of Topography (http://www.swisstopo.ch/)
- Searchable interactive map (search.ch) (http://map.search.ch/)
- Geographic data related to Switzerland (http://www.openstreetmap.org/browse/relation/51701) at OpenStreetMap

Travel

- MySwitzerland.com (http://www.myswitzerland.com/), official website of Switzerland Tourism

History

- Historical Dictionary of Switzerland (http://www.hls-dhs-dss.ch/index.php) **(German) (French) (Italian)**
- History of Switzerland (http://history-switzerland.geschichte-schweiz.ch/index.html)
- Swiss American Historical Society (http://swissamericanhistoricalsoc.org/)

Languages

- swiss-linguistics.com (http://www.swiss-linguistics.com/) Information portal on current linguistic research in Switzerland

News media

- Neue Zürcher Zeitung (http://nzz.ch/) (German), a Swiss daily newspaper
- Le Temps (http://www.letemps.ch/) (French), a Swiss daily newspaper
- Corriere Del Ticino (http://www.cdt.ch/) (Italian), a Swiss daily newspaper
- swissinfo.ch, Swiss News – Worldwide (http://www.swissinfo.ch/)
- expatinch.com, Swiss media resources page (http://expatinch.com/html/media_tv_telephony.html)
- The Local: Switzerland's news in English (http://www.thelocal.ch) Daily review of Swiss news stories in English.

Education

- Swiss Universities Handbook – Education in Switzerland (http://www.universitieshandbook.com/)
- Universities in Switzerland (http://www.swissuniversity.ch/)
- The Swiss School System (http://www.educa.ch/)
- sieps.ch (http://www.sieps.ch/) Information services on Swiss private schools and universities

Science, research, and technology

- The Swiss Portal for Research and Innovation (http://www.myscience.ch/)
- State Secretariat for Education and Research, SER (http://www.sbf.admin.ch/)
- The Swiss National Science Foundation (http://wwww.snf.ch/)

Article Sources and Contributors

Naviglio_Pavese *Source*: http://en.wikipedia.org/w/index.php?title=Naviglio_Pavese *Contributors*: EdJogg, Ian Spackman, Jllm06, Jpbowen, Moeron, Riczan, Stemonitis

Canal *Source*: http://en.wikipedia.org/w/index.php?title=Canal *Contributors*: 1122334455, 334a, A2Kafir, Ace of Spades, Ae-a, Afil, Aldis90, AlexPlante, AlexiusHoratius, Alpha Quadrant, Altermike, AlwaysUnite, Anders Torlind, Andres, Anna Lincoln, AnnaP, Antandrus, Anthony Appleyard, Anti-Nationalist, Archer7, Ares0524, Aristotle, Arpingstone, Art LaPella, Asided m plane, Aspro, Astavrou, Atif.t2, AuburnPilot, Balcer, Ballista, Balthazarduju, BarretB, Basar, BeGenderNeutral, Beachlover508g, Betacommand, Bobblewik, Bobianite, Bobo192, Brazzouk, Buffcoat, Bunthorne, Bushido Hacks, Butros, Butterflylk, CSZero, Caltas, Chase me ladies, I'm the Cavalry, Chongkian, Chris55, Christopher chalmers, Chun-hian, Clpo13, Cntras, ContinueWithCaution, Conversion script, CristiX, Curps, Cwoyte, Cybercobra, DVD R W, DVdm, Dahlis, Dakotacox1234, Dave.Dunford, David Edwards-May, DavidLevinson, Davidb0229, Deepak, Dekisugi, DerRichter, DingoUSA, Dolovis, Doron, DrDaveHPP, Drjan, Drphilharmonic, Duja, Durova, Dustimagic, Eeksypeeksy, Egw333, Elipongo, Enfantsduparadis, Engi08, Enviroboy, Erianna, Eric-Wester, Etimbo, Everyking, Ewlyahoocom, Extracareman, Ezhuttukari, Fabullus, Fifth Fish Finger, Fir0002, Fplay, G-Man, Gail, Gaius Cornelius, Gary King, Geni, Geo Swan, Gidonb, Gimboid13, Gjp23, Glenn, Glenn mc21, Gloverepp, Gman4eva, Gob Lofa, Graham87, Grahamec, Grunners, Gun Powder Ma, Gurch, Haham hanuka, HappyCamper, Hariva, HenryLi, Hertz1888, HexaChord, Hmains, Horologium, Howcheng, Hughespj, Huw Powell, Hydrogen Iodide, Hymers2, IRP, Iainscott, Iijjccoo, IronGargoyle, Ispy1981, ItsZippy, J Clear, J.delanoy, JForget, JFreeman, JaGa, JackyR, Jagged 85, JamesBWatson, Jasy jatere, Jauhienij, Jeffw57, JerryFriedman, Jh51681, Jim.henderson, Jiwa Matahari, Jmundo, JoeSmack, John0188, JohnCD, Jose77, Joshua Issac, Joy, Joyous!, Jpbowen, Kajasudhakarababu, Katieh5584, Ken Gallager, Kerry7374, Khalid Mahmood, KnowledgeOfSelf, Ksyrie, LAX, LeadSongDog, Leandrod, Leland, Lele giannoni, Leonard G., Lholt, Lightmouse, LittleOldMe, Look2See1, Luk, Lumos3, Luna Santin, Lycaon83, MCalamari, MSGJ, MacroDaemon, Magog the Ogre, Mahjongg, Mancunian101, Manea10, Marek69, Mark.murphy, Marklar2007, Markussep, Marshman, Martijn Hoekstra, Martinsach, Massimo Catarinella, Mattisse, Maxxicum, Mayalld, Mayooranathan, Melsaran, Michael Hardy, Mikhail Ryazanov, Millahnna, Miniwildebeest, Mircea cs, Mishuletz, Mongreilf, Montrealais, Mwanner, Nabla, Nakon, Narrowsword, Ndenison, Necris, Neep, NewEnglandYankee, Ng.j, Nikthestoned, Nk, Nopetro, Nposs, Nthep, Nwwaew, Nyetnyet, OGoncho, Old Moonraker, Oliver Barge, Olivier, Omnipaedista, Oxymoron83, Oydman, Parrot of Doom, Patrick, Paul W, Pennine, PericlesofAthens, Persian Poet Gal, Peter Horn, Pethan, PhnomPencil, Piano4444, Piglet, Pigsonthewing, Pilgaard, Pinethicket, Pradeepsomani, Prashanthns, PrestonH, Primasz, Proabivouac, Pyrotec, Qbmessiah, Quadell, RDL, RJaguar3, Ranty Burns, Raso mk, Regregex, Renata, Resinguy, RexNL, Rich Farmbrough, Riczan, Rjensen, Rjwilmsi, Robertvan1, Rodw, Roquai, Rude2o2, SDC, SEWilco, Safemariner, Salimfadhley, Sandstein, Sc3499a, Shannock9, Signalhead, Silivrenion, Silverchemist, Singhalawap, Some jerk on the Internet, Specs112, Stephenb, Stevo1000, Stubblyhead, Stwalkerster, Svick, Sweetness46, SwinSvinoza, Symane, TYelliot, TakuyaMurata, The very model of a minor general, Thehelpfulone, Themepark, Tholly, Tide rolls, Tomtom08, Tonyle, Trist3n, Ttony21, Turk oğlan, TutterMouse, Ultimus, Unschool, Vald, Vera.tetrix, Vicki Rosenzweig, Viriditas, VirtualSteve, Viskonsas, Watinf, Wavelength, Wayne Slam, Wetman, Wikipedia is communism 2, Wimt, Wlodzimierz, Woohookitty, Wordbuilder, XJamRastafire, Yamamoto Ichiro, Yintan, Zeisseng, Zuejay, ميلان لىقع فشا, 451 anonymous edits

Navigli *Source*: http://en.wikipedia.org/w/index.php?title=Navigli *Contributors*: EdJogg, Ian Spackman, Ita140188, Jllm06, Melaen, Ms2150, Northfox, Riczan, Stemonitis, 4 anonymous edits

Lombardy *Source*: http://en.wikipedia.org/w/index.php?title=Lombardy *Contributors*: *Paul*, 2deseptiembre, A purple wikiuser, AV3000, Abacos, Adashiel, Agne27, Ahoerstemeier, Aid85, Alan Liefting, Aldux, Alexander Domanda, Alexwcovington, Andre Engels, Angelo.romano, Angr, Annapavoni, Argo Navis, Arthena, AstroNomer, Attilios, Avoided, Barticus88, Beetstra, Berek, BernhardtP, Bill Thayer, Borgerone, Broletto, Bullzeye, Bum hole, Bunnyhop11, Caerwine, Calmer Waters, Caponer, Carlossuarez46, Cavarrone, Chanheigeorge, Checco, Chickyfuzz, ClickRick, Clutcher, Cncs wikipedia, Codice1000.en, Commander Shepard, CommonsDelinker, Conte di Cavour, Crisco 1492, Cunibertus, Czolgolz, D C McJonathan, D6, Dakrismeno, Daufer, David Kernow, DerBorg, Deusdemona, Dia^, Dimitrii, Dolphonia, Doug, Downloadmeh, Dreadstar, Dsp13, Eboracum, Ed g2s, El C, Elapsed, EliasAlucard, Emilia-Romagna, EnSamulili, Enrico 69, Epbr123, Erianna, Everyking, Explicit, F l a n k e r, Fertuno, Fiet Nam, Folini, Fred Bradstadt, Frokor, Gabstef, Gaius Cornelius, Gazilion, Gennarous, Giddylake, Giggity120, Giorgio, Gjd001, GraemeL, Grafen, Grammarmonger, Grifter72, Grstain, Gryffindor, Hadal, HalfShadow, Hayden120, Helloimcraziiii, Himasaram, HisSpaceResearch, Historymike, Hotcrocodile, Ian P Walker, Ian Spackman, Ian Stockdale, Isaaccates, Italianparks, Ivirivi00, Iwillremembermypassthistime, J.T.Rossi, JCRansom, JLogan, Jac16888, Jake.panozzo, Jauhienij, JdeJ, Jeronimo, JinJian, Joefran4, John5Russell3Finley, Johnluisocasio, JonC0001, Jose77, Joseph Solis in Australia, Kevin, Kintetsubuffalo, Knorrepoes, Kostisl, Kricxjo, Kumo, Kwamikagami, Kweedado eng, LilHelpa, Lincher, Louisbeta, LtNOWIS, Luc19, Lucifero4, M7, Magioladitis, Mai-Sachme, Malcolm Farmer, Maokart444, Marco Bonavoglia, Marcomob, Marek69, Markussep, Matijap, Matwi, Meddlin' Pedant, Melaen, Merchbow, Mervyn, Mic, MichaelTinkler, Midnight Green, Mild Bill Hiccup, Mindmatrix, Mintleaf, Ms2150, Mushroom, Mwanner, Mygerardromance, MysteryHeff?, Nero the second, Neutrality, Nicola.Manini, Night of the Big Wind Turbo, Nimrod88, Nineko, Ninonino, Oghmoir, OllieFury, Open2universe, Oreo Priest, Ortcutt, Paolotacchi, Paul August, Peter.C, Pfalstad, Piccolo Modificatore Laborioso, PoccilScript, Putney Bridge, R'n'B, Radon210, RainbowCrane, Rarelibra, Raven in Orbit, Rdsmith4, RingtailedFox, Rl, Rmhermen, Robert K S, Romanm, Rrenner, Rune.welsh, Samoano, Sandstein, Saruwine, SchfiftyThree, Sdoerr, Shadowjams, Shawnhath, Shuipzv3, Skinsmoke, Snoyes, Stan Shebs, Stemonitis, Str1977, TXiKi, Tabletop, TangoItalia, Template namespace initialisation script, The wiki editer master, Theologiae, Tigershrike, Tommy2010, Tony1, Trekphiler, Tzetzes, Ugur Basak, Uogl, Vega84, Vicki Rosenzweig, Virus326, Wai Hong, Walden69, Wavelength, Wetman, Why Not A Duck, Wiki alf, Woohookitty, Worthawholebean, Yegorius, Yorkiano, தகவலுழவன், 473 anonymous edits

Italy *Source*: http://en.wikipedia.org/w/index.php?title=Italy *Contributors*: (GC)KYE, *Kat*, -- April, -OOPSIE-, -kayac71-, 000peter, 0lorenzo0, 100110100, 100pfanner, 1234r00t, 151.24.188.xxx, 151.24.190.xxx, 23prootie, 24.218.142.xxx, 2812, 28421u2232nfenfcenc, 2mrchio, 2t56, 334a, 55hh44gg, 72Dino, 7852, 8k, 97198, A Werewolf, A2Kafir, AKeen, ALargeElk, AMK390, ANGELUS, AP1787, APayan, ARC Gritt, Aaron Einstein, Abcd24, Abonazzi, Achangeisasgoodasa, Ackander, Actarux, AdRock, Adam Bishop, Adashiel, Adavis444, Addshore, Adrecaled, Adrian, Aelwyn, Afterwriting, AgentCDE, Ahdtravel, Ahmad123987, Ahoerstemeier, Aid85, Aigos100, Aitias, Aivazovsky, Akamad, Akanemoto, Akbg, Al bello al, Alabamagene, Alagemo, Alarm, Alasdair, Alcaffemoro, Aldux, Alec74, AlefZet, AlenWatters, Alessandro57, Alex Bakharev, Alex S, Alex inmot, Alexander Domanda, AlexanderFreud, Alfabookings, Alfio, Alib, Alien life form, Alienoster, Alinor, Alison9, AllyUnion, Alphaboi867, Alphachimp, Alphathon, Alv21, Amakuru, Amazins490, Amcfreely, America9, Amylyn, Ana570, Analytikone, Andrea Censi, Andrea Cioffi, Andrew Hampe, AndrewHowse, Andrewpmk, Andy Marchbanks, Andy120290, AngelOfSadness, Angela, Angelo De La Paz, Angelo Somaschini, Angelo.romano, Anger22, Animum, Anonymous editor, Antandrus, Anthere, Anthony.bib, Aogouguo, Aowpr, Aplaso, Applesrgood99, Aranel, Aratuk, Aremith, Aris Katsaris, Army935, Arnaldo Mauri, Arod14, Arpingstone, Arronax50, Art LaPella, Art10, Arthir7, Artichoker, Ashdod, Ashkani, Aside, Asnac, Astronaut, Attilios, Atwardow, AuburnPilot, Aussinator, Australian cowboy, Avala, Avatarion, Avaya1, Avenue X at Cicero, AvicAWB, Avicennasis, Avij, Avillia, Avionics1980, Avnjay, Avsoccerplanet94, AxG, Axeman89, Aytharn, B. Phillips, BGManofID, BNewman, BRG, Badseed, Bake says hi, BalticPat22, Bambuway, Banes, BanyanTree, Barbaricino, Bardhylius, Baristarim, Barneca, Baronnet, Barryob, Bart133, Baruneju, Basicdesign, Bassem18, Battem, Battleagainstbald, Bazonka, BazookaJoe, Bbartlog, Bbb1992, Bbsupdee, Beardo, Beetstra, Before My Ken, Beginning, Begoon, Beland, Belligero, BenWoolley, Bender235, Bentley4, Berek, BernardaAlba, Besidesamiracle, Beta m, Betacommand, Bevo74, Beyond silence, Bgs022, Bhadani, Big Adamsky, BigBen21, BilCat, Bill Thayer, Binksternet, Birdman1, Bkcraft, Bkell, Bkivhtir, Black Falcon, Black Kite, Black sheep997, BlackLukes, Blake24, Blehfu, Blindskater1234, Blinutne, Blizzardstep0, Blocci, Blue-Haired Lawyer, Blue520, Bluelion, Bluemask, Bob.v.R, Bobblewik, Bobbyrullo, Bobbytheimpaler, Bobo192, Bodie282, BogdaNz, Bogdangiusca, Bonaparte, Bongwarrior, Bookofjude, Boothy443, Boraczek, Borameer, Borgerone, BovineBeast, Bpollock510, Bradallen, Bradenripple, Brainboy109, Branddobbe, Brandmeister (old), Brandonleemchugh, BrendelSignature, BrendonJS, Brian Pearson, Brian0918, BrianGV, Brianga, Brion VIBBER, Brutaldeluxe, Brutannica, Bryan Derksen, BryanG, Bthayesesq, Buaidh, Bubba hotep, Bucketsarehuge, Bucketsofg, Buddy375, Bukkia, Bulgariasucks, Bull Market, Bullet83, Bullshark44, Bully25, Burndownthedisco, Burntsauce, Burto88, CALR, CJ Aikman, CRE is king, Cactus.man, Cacycle, Cadsuane Melaidhrin, CaesarAugustusDivusIulius, Caesarjbsquitti, Cal Burrattino, CalJW, Calamitas-92, Calaschysm, Calculuslover, CalicoCatLover, Caltas, Calton, CambridgeBayWeather, Camembert, Campolongo, Can't sleep, clown will eat me, CanOfWorms, CanadianLinuxUser, Canderson7, Caniago, CanisRufus, Cannam, Canterbury Tail, Canthusus, Cantus, Carl.bunderson, Carlobortone, Casper2k3, Cassiodorus, Catalyst in Society, CatherineMunro, CattleGirl, Caucasicus, Cavrdg, Cbrown1023, Cburnett, Ccarroll, Ccirulli, Cdc, Celestra, Celsius1414, Centrx, Ceyockey, Charbear08, Charles Matthews, CharlesMartel, Charlir91, CharlotteWebb, Charvex, Chasdenonno, Che829, Checco, Chelsland, Children of the dragon, Chkua, Chris the speller, ChrisGlew, Chrishmt0423, Chrisieboy, Christoff2k7, Christopher Kraus, Christopher Mann McKay, Christopher Parham, Chuckiekiller, Chucknorrisfan, Cibjibb, CieloEstrellado, Civil Engineer III, Ck lostsword, Ckatz, Clacsatron, ClamDip, Clap, Clarkcj12, Clem cowsie, Clngre, ClockworkTroll, Clyde Miller, Cobblet, Cocoaguy, Cody12343, CodyJoeBibby, Cogito ergo sumo, ColdShine, Colonel Marksman, Colonies Chris, Commander Keane, CommonsDelinker, Conaj07, Conglacio, Connolly15, Conone, Conte di Cavour, Conti, Conversion script, Conwatts, Cool3, Coolhawks88, Coredesat, Corpx, Corti, Cratbro, Crausch932, Crazycomputers, Crazyncfan1, Crazytales, Creedence, Creidieki, Cremepuff222, Cricket93, Crispus, Cristianaladki, Crowish, Crusadeonilliteracy, Crònica, Cs-wolves, Css, Ctu2485, Cunibertus, Curps, Cusio, Cuttie+hottie, Cuvtixo, Cvene64, Cyanolinguophile, Cyberchimp, Cybercobra, Cyclonenim, D4niel11, D6, DARTH SIDIOUS 2, DCB4W, DGJM, DJ Tricky86, DRTllbrg, DSuser, DTOx, DVD R W, Da biGGeSt ItaLiANo yo, Daddydavide, Dakrismeno, Damon Mah, Dandan44sou, Dangerous-Boy, Daniel Olsen, Daniel5127, Daniel563, DanielCD, Danieldanielnova, Danno, Danny, Dannyc77, Dantadd, Dantetheperuvian, Danthemankhan, DanyaRomulus, Dark Shikari, Darkchaos556, Darkoneko, Darth Vadir, Daveb2k1, Davewild, Daviboz, David Corbett, David Gerard, David Kernow, David Liuzzo, David Newland, David Parker, David Underdown, DavidRF, Davide41, Davidok, Daygum, Dbachmann, Dbenbenn, Dbrodbeck, Dbtfz, Dchall1, De Administrando Imperio, Debresser, Decay 2, Deco16-10, Deflective, Degrak, Deibid, Deleting Unnecessary Words, Delirium, Delldot, Dengero, Dep. Garcia, DerHexer, DerMeister, Derek Ross, Derek's A Snorer, Derekwriter, Desiphral, Desmond Tsui, Deville, Dewrad, Dexippus, Dfrg.msc, Dharmabum420, Dhiresh, Did01010101, Digitalme, DiiCinta, Dillonstars, Dimitrii, Dina, Dirtyfrank10, Disavian, Discospinster, DivineIntervention, Diwas, Dizzle101, Dlohcierekim, Dmmaus, DocSigma, Docu, Dodoïste, Domaleixo, Domino theory, Donrub, Donutman7, Dorftrottel, Dori, Dougweller, Dr. Henri Monet, DrBob, Dreadstar, Drmies, Drubakawaski, Drugonot, Duckswarrior8, Duomodimilano, Durruti, Dussst, Dustimagic, Dwayne, Dynaflow, Dysepsion, E Pluribus Anthony, E vattimo, E-boy, ESkog, EaglesFanInTampa, Eaomatrix, EazieCheeze, Echuck215, Eclecticology, Economichostels, Editor18, Editor88, Edivorce, Edward, Eeamoscopolecrushuva, EgraS, Eirik Raude, Ejk81, El C, El Greco, ElSaxo, Elastix DePilus, Eleassar, Electionworld, Elemental Fury, Elf-friend, ElinorD, Elkman, Elliskev, Elm-39, Elmondo21st, Elockid, ElockidAlternate, Eltener, Elzippo, Ema92, Emberstone777, Emc2, Emersoni, Emes, Enchanter, EncycloPetey, Energetic is francine@yahoo.com, Enkyklios, Enok, Enric Naval, Entirelybs, Enviroboy, Enzino, EoGuy, Epbr123, Epf, Epischedda, Equilibrial, Ercole, Erinaceus, Ertz, Esimal, Espo111, Estoniaee, Estoy Aquí, Ettormo, Etz Haim, Eu.stefan, Eugene van der Pijll, Ev, Evercat, Everwillow, Everyking, Evlekis, Example111, Excaladbiur, Excirial, Exert, Experia, Explendido Rocha, Exult, FDAU, FF2010, FRANKENSTONED, Fabhcún, Fabricationary, Factanista, FactionofReds, Faithlessthewonderboy, Fang Aili, Fantasy, Fantomf1, Farenheit451, Farkas János, Farside6, FastLizard4, Fatbad, Favonian, FayssalF, Fbergo, Fbv65edel, Fds j a, Felipe Menegaz, Felisopus, Ferro Carlotta Monzi Brak, Fertuno, FetchcommsAWB, Fieldday-sunday, Filipdr, Fireaxe888, Firehawk1717, Fiscap, FisherQueen, Flabreque, Flamey182, Flatterworld, Fleaman5000, Flibjib8, Flyguy649, Fmon, Fncfanyoyo, Focak, Folks at 137, Foscoe, Foundert, FoxCE, Fram, France91449144, FrancisTyers, Francomemoria,

Francs2000, Franjklogos, Frazzydee, Freakofnurture, Freddyballo, Fredrik, Freepenguin, Fremsley, FreplySpang, Frietjes, Frymaster, Funnyhat, Future Perfect at Sunrise, Fuzheado, G. Campbell, G.-M. Cupertino, GDonato, GFlusitania, GGbonazzi, GHe, GKoUtSo201, Gabbe, Gadfium, Gadget850, Gaff, Gaius Cornelius, Galati, Galloramenu, Galoubet, Ganeshk, Garfield226, Garlics82, Garretgian, Garry Saint, Esquire, Garzo, Gator1, Gaz, Gbaddorf, Gbnogkfs, Gdo01, Gene s, Geni, Gennarous, GeoWriter, Gerry Lynch, Gfoley4, Gggh, Ghepeu, Ghosts&empties, Gianfranco, Giddylake, Gidonb, Giespe, Gilgamesh, Gilliam, Gimmetrow, Giordaano, Gioto, Giovanni33, Giuggiola90, Giuliof, Giving is good, as long as you're getting, Glacious, Glen, Glenn, Gmatsuda, Gnevin, Gnfnrf, Gogo Dodo, Golbez, GoldFlower, Goldiegopher, Goochelaar, Good Intentions, Good Olfactory, Goustien, Gracenotes, GraemeL, Grafen, Graham87, Gramaic, Gravitan, Grblade, Green Giant, Green Owl, GreenLocust, Greenmars, Gregorik, Gresszilla, Grifter72, Ground Zero, Gryffindor, Gsarwa, Gubbubu, Guettarda, Gugganij, Guilherme Paula, Gurch, Gurchzilla, Guy Peters, Gwernol, Gzornenplatz, HJ Mitchell, HVH, Haakon, Habs4ever, Hadal, Haemo, Hahaha12345, Haham hanuka, Hajor, HangingCurve, Harry R, Hashem sfarim, Haukurth, Hayabusa future, Hayden120, Hdt83, Hellisp, Helsinkicommission, Henry Flower, Henrygb, Henrymark, Hermione is a dude, Herostratus, Hibernian, Highvale, Hirond, HisSpaceResearch, Historian19, Hitec81, Hjncfkdnmhbjk, Hlanefaber, Hmains, Hmusseau, Hogyn Lleol, Honette, Hotlorp, Hu12, Hullaballoo Wolfowitz, Human Rights Believer, Hunfe, Huon, Hurricane111, Husond, Hut 8.5, II MusLiM HyBRiD II, IRP, IRelayer, Iain99, Ian Spackman, Ianb, Icairns, Icsunonove, Idaho spud, Idaltu, Ief, Igatsios, Ikkyu2, Il Moderato, Ilkeston1990, Imgay, Imperfection, InShaneee, Incredible007, Indon, Inge-Lyubov, Inner Earth, Int21h, Intelati, Inter, Inter16, Interchange88, Ioannes Tzimiskes, Ionut.shaggy, IrfanFaiz, Iridescent, Irishguy, IronGargoyle, Ironplay, Irrer, Irrypride, Islamablad, Ispy1981, It.wiki:Twilight, Itaguy, Italfan, Italian boy, Italianfever, Italianguy1, Italiano111, Italyandus, Ithanyx, Ithinkhelikesit, Ivirivi00, Iwanttoeditthissh, Ixfd64, J Di, J Milburn, J hayter, J.K Rowling, J.Winston, J.delanoy, J04n, J0mb, J36miles, JEN9841, JForget, JFreeman, JHMM13, JJJJust, JLogan, JNW, JTBX, JYolkowski, JaGa, Jaan, Jacek Kendysz, JackSlack, Jackofh3arts, Jacob Haller, Jaeggerlover, Jahsonic, James miuccio, JamesR, Jamie C, Janneman, Jarble, Javit, Jayden54, Jazriel, Jcr2, Jcro76, Jcw69, JdeJ, Jeanambr, Jebba, Jedi tales, Jensboot, Jeronimo, Jerryseinfeld, Jersyko, JesusSavesNarnia, Jewfacejared, Jhendin, Jiang, Jimtaip, Jinxed, Jj137, Jjansen1234, Jjhcap99, Jjjsixsix, Jklin, Jlittlet, Jmabel, Jmlk17, Jnestorius, JoLoB, JoSePh, JoanneB, JoeBlogsDord, Joffeloff, John, John of Reading, John254, Johnhardcastle, Johnleemk, Jonathan.s.kt, Jonik, Jono1111, Jonpro, Jordan Elder, Jorunn, Jose77, Joseph Solis in Australia, JoshHolloway, Joshua, Jossi, Joy, Jpeob, Jrinaldi, Jsc1973, Jsdixie uk, Jstiller, Jszabo, Julia L., Jumbuck, Junior Brian, Junyor, Jusdafax, Jusjih, Justlettersandnumbers, Juve2000, Jvian, Jytdog, Jza84, K.d.stauffer, KJS77, KJohansson, Kaasje, Kablammo, Kaihsu, Kaio-ken x10, Kalathalan, Kalrotenmw, Kansoku, Karimarie, Karl, Karmafist, Kaspo, KathrynLybarger, Katij13, Kaye28, Kbh3rd, Kbolino, Kcops7, Keegan, Keegansmells, Kenshin, Kevin, Kevin B12, Kevin Murray, Kevo-723, Kfrharris, Kgrad, Khazar, Khoikhoi, Kiensvay, Kimchi.sg, King of Hearts, King wiston, Kingston28, Kinneyboy90, Kintetsubuffalo, Kjwwww9, Kkm010, Kman543210, Kmusser, Knidu, Knowledge Seeker, KnowledgeOfSelf, Konstable, Korg, Kotniski, Kozuch, Kr1st1 crazydeejay97, Kr1st1deejay97, Krakatoa, Krawi, Krich, Kristenq, Krun, Ksenon, Kubigula, Kugarlad, Kungfuadam, Kungming2, Kuru, Kusma, Kvdveer, Kwame Nkrumah, Kwamikagami, Kwertii, Kyle Andrew Brown, Kyle sb, Kyleisme, KymeSnake, L 0 0 P, L Kensington, LMoffat, LOL, LOM, LRT24, Lacrimosus, Lagalag, Lahiru k, Lakers, Langston, Lazylaces, Leafyplant, Lear 21, LeaveSleaves, Lectonar, Lefairh, Lemonflash, LeoO3, Leon..., Leonardomio, Lesfer, Leslie Mateus, Lethalreno, Levineps, Lewis R, Lexicon, Lfh, Lhoaxt, Liftarn, Ligguma, Lightdarkness, Lightmouse, Lights, Ligulem, LilHelpa, Lilac Soul, LinkGT, Linkspamremover, Linnell, Linuxbeak, Liox8, Lip zort, Liquidmetalrob, LittleJerry, Ljconway, Lkinkade, Llull, LlywelynII, Lockesdonkey, Lol man, Lollipop Lady, LonelyMarble, Lonwolve, Looris, Looxix, Lord Palmerston, Lord Vader, Lord Voldemort, Lordkazan, Lorenzino, Lornova, Lradrama, Luc19, Lucas323, Lucifero4, Luisztdt, Luk, Luna Santin, Lunkwill, Lupin, Lupo, Luqui, Lussier, M412k, MBE, MER-C, MGTom, MJCdetroit, MONGO, MZMcBride, Mablin76, Mac Gille Domhnaich, Macedonian, Machadoman, Macrakis, Madmonk325, Magicmonster, Magioladitis, Magister Mathematicae, Mai-Sachme, Mailer diablo, Maiprost, Majorly, Malarious, Malcolm Farmer, Malhonen, Mallerd, Malo, Malwinder25, Man vyi, Mana Excalibur, Mantellg, Marco Neves, Marco Plassio, Marco bisello, Marcos G. Tusar, Marcus2, Marcuscalabresus, Marcworca, Mareino, Marek69, Marginoferror, Mariokempes, Mark, MarkGallagher, MarkusHagenlocher, Markusavedal, Marlowe, Marnanel, Martinwilke1980, Matchrthom, Materazzi23, Materialscientist, Matt Crypto, Matt1001, Mattie123456, Mattiecat, Matwi, Mav, Maxallen, Maximus Rex, Maxí, Maybe123, Mbimmler, McDogm, Mcarroll63, Mck0009, Mcnealy, Me.play, Me4321, Mebizzare, Mediterraneo, Meegs, Megan1967, Mel xox, Melsaran, Memetician, MercZ, Merovingian, Mervyn, Mesgul82, Mets501, Mfa fariz, Mgiganteus1, Mgw, Mhking, Mic, Michael Hardy, Mickey gfss2007, Mickeykahn, Mike Rosoft, Mike Selinker, Mikel506, Mild Bill Hiccup, Milkbreath, Millisits, Millosh, Mimmo46, Minaturekobe08, Mindmatrix, Mindspillage, Minesweeper, Minna Sora no Shita, MinnesotanConfederacy, Miranda, Miskin, Miss Madeline, Missmarple, MistyMorn, Mkpumphrey, Mlaffs, Mmhrycak, Mo0, Moelarry, Moncubus, Monkbel, Monkeygurl12, Montrealais, Moonraker, Moreschi, MorganaFiolett, Moriori, Morrad, Morwen, MosheA, Moverton, Moxy, Mr italia, Mr. G. Williams, Mr. Yooper, MrPMonday, Mrjrdnthms, MrsWeasley&&Dracoscousin, Mtaylor848, Muad, Mushroom, Mvphothits8849, Mwanner, Mwilso24, Mxcatania, Mxn, Mydaz, Mynabird, Mysidia, Mzbello, NEMT, NHBaritone, NHRHS2010, Nabilus junius, Nabla, Nach0king, Nafis ru, Naive cynic, Nakon, Naohiro19, Nascimbeni, Natalie Erin, NatusRoma, Navisliburnia, NawlinWiki, Nazare, Nbowers10, Ndn, Necatrix Regina, Nehrams2020, Nehwyn, Neilc, Nekka, NetherlandishYankee, Netoholic, Neutrality, NewEnglandYankee, Nichalp, Nick Linsmayer, Nick.mon, Nick543, Nickel Chromo, Nicksukh, Nico, Nicolino22, NigelR, Nighthealer, Nightkey, Nightstallion, Nigosh, Nihil, Nikai, Nikknighten, Niko87, Nikodemos, Nishkid64, Nitya Dharma, Nivix, Nlu, Nmpenguin, Nnemo, NoIdeaNick, NoSeptember, Nobleeagle, Nochoje, Noclador, Nofrillz, Nog64, Nono64, Noozgroop, Norrild, NorsemanII, Northumbrian, Notafly, Notcarlos, Nothing149, Nova77, Novalis, Nowyouseeme, Npovshark, Nrbelex, Nsaa, Nskrill, Nthornton, Ntsimp, NuclearVacuum, NuclearWarfare, Number 0, Numberonesurvivor, Numerousfalx, Nyttend, OMGsplosion, OOODDD, Obenaza, Obli, Oblivious, Obradovic Goran, Ocaasi, Ochristi, Oda Mari, Ohconfucius, Ohnoitsjamie, Okapi, Oktawiusz, Olazzo, Olegwiki, Oli Filth, Oligopolistic, Olivier, Olorin28, Omicronpersei8, Onco p53, OneGuy, Onetwo1, Oo64eva, Opelio, Opinoso, Optiguy54, Optimist on the run, OrbitOne, Oreo Priest, Ornitorinko, Oxymoron83, Oyabun83, Oyvind, Ozzman commeth, PAK Man, PDH, PIO, PIrish, PZFUN, Pabix, Pablo-flores, Pacedm, Paine Ellsworth, PalumboRoberto, Pamri, Panairjdde, Panarjedde, Pani0202, Panser Born, Paolo da Reggio, Paolotacchi, Pap3rinik, Pascar, Pasquale, Pastricide, Patavium, PaterMcFly, Patricknoddy, PatrikR, Paul August, Paulinho28, Paulista01, Pavel Vozenilek, Pax:Vobiscum, PaxEquilibrium, Pb30, Pearle, Peeperman, Pegasus1138, Pemble, Penny babi, Pensionero, PerfectStorm, PericlesofAthens, Persian Poet Gal, Persianne, Perullama, Peter, Peter Horn, Peterdjones, PeterisP, Pethr, Pevarnj, Pgk, Phaedriel, Philip Trueman, Philip200291, PhilipC, Philopoemen, Phucmai, Phunting, Picapica, Piccolo Modificatore Laborioso, Pierpaolo.Dondio, PierreAbbat, Pietro, Pilotguy, PinchasC, Pinethicket, Pirawik, Pit, Piutus, Plastikspork, Pmadrid, Poindexter Propellerhead, Polaron, Polemyx, Polyphilo, Ponpan, Poppa Yami, Porqin, Potito m. petrone, Praveenp, PrestonH, Prestonmcconkie, PrimeNumber, PrincessofLlyr, Prodego, Profitoftruth85, Prolog, Psy guy, Pudeo, Puertorico1, Purpleturple, Putdust, Pwqn, PyroBlaze272, Python eggs, Pádraic MacUidhir, Qqwwwwwwww, Qtoktok, Quaeler, Quantpole, Quarma, Quendus, Quiensabe, Quizimodo, Quizimodo (usurped), Qwe, Qwyrxian, R, R Lowry, R'n'B, R-41, R9tgokunks, RBrez, REAL MADRID, RG2, RVD, Radarino, RagingR2, Rama's Arrow, Ramirez72, Ran, Random user 39849958, Rarelibra, Raspberrysnapple, Raven in Orbit, Raymond arritt, Rdonnelly212, Rdsmith4, Realek, Reckrap, Reconfirmer, Recurring dreams, Redking7, Redvers, Regara MkII, Reisio, Reo On, Retiono Virginian, Rettetast, RexNL, Rhialto, Rholton, Ribbygoose, Rich Farmbrough, Richardcavell, Rick Block, RickK, RickyTheBest, Riczan, Rikimaru, Rjwilmsi, Rkoboy15, Rlest, Rmeht, RoadTrain, Rob.derosa, Robekun Mengoni, RobertG, RobertM525, Robster1983, Rock2e, Rocker321, Roke, Roleplayer, Romaioi, Romanm, Ronjr708, Ronline, Rosh2610, Rospo Matto, Rotring, Roux, Roy Al Blue, RoyBoy, Royalguard11, Rror, Ruchi tspl, Rudyisaho, Rumpelstiltskin223, Rupertsciamenna, Rursus, Rustyfence, Ryan Postlethwaite, Ryulong, SBKT, SDMMC, SJP, SLV100, SLi, SMC89, SSJ4 Aragorn, SSZ, SUP.TRO., SWAdair, Sacchipersempr, Sacristán, Sadistik, Sailko, Salimzaatari, Salvo46, SalvoCalcio, Sam Hocevar, Samantha555, Samfreed, Samir, Samoano, Samrolken, Samuel, Samw, Sango123, Sanmartin, Sara p1414, Sarah, Sardanaphalus, SarekOfVulcan, Savie Kumara, Scanos, Sceptre, SchfiftyThree, Schubatis1, Schzmo, Scientizzle, Scipius, Scitzophrenic, Scj2315, Scohoust, Scottbell, Scouller, Scoutstr295, Scruffy brit, Scythian1, Sdoerr, Sean.hoyland, Sebbex, Secfan, Seeker alpha806, Semmler, Semperf, Senator Palpatine, Serenesoulnyc, Seresin, SeriousUser, Serpent's Choice, Serte, Sethflorida2007, Seurope, Sfair, Sgt.fox, Shadowjams, Shanel, Shanes, Shannon bohle, Shardan, Shawnhath, Shell Kinney, Shenme, Shenron, Shichinintailuv, Shizane, Shoaler, Shoeofdeath, Shoessss, Shredder2012, Shroudan, Sicilianmandolin, Sideshow Bob Roberts, Signalhead, Signsolid, Silbertanne3, Silly rabbit, Simbamford, Simetrical, SimonP, Simone, SineWave, Sintaku, Sir John Alexander Macdonald, Sir Nicholas de Mimsy-Porpington, SirEbenezer, Sjakkalle, Skb8721, Skizzik, SkyBoxx, Skäpperöd, Slakr, Slark, Slasheh, Slater bob, SleepyDan, Slideshow Bob, Slowking Man, Sluzzelin, Smalljim, SmartGuy, Smg 777, SmthManly, Snbritishfreak, Snowmanradio, Snoyes, SomeHuman, Soosed, Sophrosune, Soprani, Sortior, Souris2005, SpNeo, Spacedragonblue, SparrowsWing, Spartan-James, Spartan15DEMO, Special-T, Spettro9, Spliffy, SpookyMulder, Springeragh, Spud Gun, SpuriousQ, SquallLeonhart ITA, Srikeit, Srose, Sroulik, Ssbohio, Ssolbergj, Staffwaterboy, Stan Shebs, Standfest, Stardust8212, Staringatmygrave, Staxringold, Steel, Stefano.nicolucci, Stefo, Stemonitis, Stephen G. Brown, Stephenb, Stephreef, Stereotek, Sterlsilver, Steven Weston, Stewie814, StoneProphet, Subtropical-man, Suhardian, Sunquanliangxiuhao, Sunray, Superbeecat, Supparluca, Surf R Dude, Suryoye, Svick, Swarve, Swid, Sygcom, Sylvain1972, Symplectic Map, Syrthiss, Syvanen, T 587, TBadger, TFNorman, THEN WHO WAS PHONE?, THe ZiPMaN, TIM Lewis ACS, TOttenville8, TUF-KAT, TaalVerbeteraar, Taalo, Tabletop, Tabortiger, Tachin34, Tadpole-Rulz, TagarSpinebreak, Tagishsimon, Talon Artaine, TalonGold, Tangotango, Targeman, TarquiniusWikipedius, Tasc, Tassedethe, Taueres, Taulant23, Taximes, Tbhotch, Tburingham, Teddycom, Teeberry, TehranIran, Tekleni, Telex, Terence, Terfili, Terra Novus, Tesaux, TexasAndroid, Thanneer, Thatdog, Thayts, The Anome, The Kalibos, The Nixinator, The Rambling Man, The Transhumanist, The tooth, The undertow, TheKMan, TheOne12345, TheOne123455, TheST, TheSpook, TheTrojanHought, Thebass66, Thebluebeast, Theda, Theirrulez, Theitalianboy, Theo van der Zalm, Theologiae, Therefore, Therequiembellishere, Therevan, Thewallowmaker, Thewikipedian, Think outside the box, ThinkBlue, Thisisbossi, ThomasK, Thome66, Thricecube, Thunderbird2, Tiddly Tom, Tide rolls, Tifego, Tiger888, TigerShark, Tigging, Tim riley, TimComm, Timberframe, Timc, Timir2, Timo Laine, Timwi, Titoxd, Tjl1128, Tktru, Tobby72, Tobes00, Tobias Conradi, Tom harrison, Tombelgrano, Tombomp, Tomeasy, Tomi, Tommaso.gavioli, Tomtheman5, Tomtom1156, Tomwalden, Tonalone, Tone, Tontotti, Tony1, TopoConvention, Tourtux9, TownDown, Tpbradbury, TptmasterHalifax, Tr6637, Trangshi, TransUtopian, TreasuryTag, Trevor MacInnis, Trigaranus, Trompeta, Trovatore, Trusilver, Tucri, Turkish Flame, Turkuun, Tuspm, TwoOneTwo, Tyler, Typ932, Tyrol5, Tzartzam, U.Steele, UAAC, Uanabana, Uccello di Fenice, Ufwuct, Ugur Basak, Ulmanor, UltimaRatio, Ultimate Destiny, Ultranol, Ulyaoth, Un chien andalou, Un sogno modesto, Ungvichian, UnicornTapestry, Unyoyega, Uomodis08, UpstateNYer, Us441, Useight, UserDoe, Usergreatpower, Utcursch, V6g3h7, VKing, Valenciano, Valentinejoesmith, Valentinian, Van helsing, Vaniac, Vaniba12, Vanished 6551232, Vanished user 03, Vardion, Variant, Vdjj1960, Vega84, VegaDark, Vegaswikian, Veledan, Vendrov, Ventur, Venu62, Verdadero, Verrai, VeryVerily, Viator slovenicus, Viewfinder, Vikkoz, Vilallonga, Vincenzo, Vinsfan368, Vinyanov, Violncello, Virgule82, VitoScaletta, Vivaitalia57, Vivio Testarossa, Vkem, VladB, VolatileChemical, Voldemort, Voldemort175, Vorik111, Vsmith, W34org, WHATaintNOcountryIeverHEARDofDOtheySPEAKenglishINwhat, WJBscribe, WTGDMan1986, WTucker, Wahrhaft, Wai Hong, Walsh7, Walter Görlitz, Waltloc, Wangi, WarFox, WarraneIsGay, Washdivad, Washingm2, Waswaswas111, Wavelength, WegianWarrior, Welsh, Wento, Weregerbil, Werson, West Brom 4ever, Wetman, Weyes, Wguynes, Whaatt, Where, WhisperToMe, Whispering, Whitedwarf1.3, Whomp, WikHead, Wiki alf, WikiLaurent, Wikianon, Wikiborg, Wikigregor, Wikihistorian, Wikipedia Administration, Wikipedius, WildWildBil, Will Beback Auto, Willsamuel, Willstu, Wimt, Winhunter, Wino911, Wisden17, Wknight94, Wlmg, Wolfer68, Woohookitty, WookieInHeat, Wwoods, XMillar, Xain it, Xaosflux, Xevious, Xezbeth, Xnuala, Xuxo, XxSavannahxX, YUL89YYZ, Yahel Guhan, Yamamoto Ichiro, Yanksox, Ybbor, Yc183, YellowMonkey, Yelyos, Yemal, Yoninah, Yorkshirian, Yousou, Yousuck101, Yuckfoo, Zacharie Grossen, Zap Rowsdower, Zargon2010, Zhouf12, Zidanie5, Zimbricchio, Zisa, Zocky, Zoe, Zondor, Zoney, Zro, Zscout370, Zsinj, Zundark, Zzuuzz, Zzzzzzus, Žiedas, ΚΕΚΡΩΨ, קרע, 4173 anonymous edits

Ticino_(river) *Source*: http://en.wikipedia.org/w/index.php?title=Ticino_%28river%29 *Contributors*: 8loginout8, CG janitor, Caspian blue, Cate, Charles01, Corti, Dakrismeno, Darwinek, Docu, El C, Frietjes, Frokor, Ggonnell, Gilliam, HennessyC, Ian Spackman, Iohannes Animosus, J.P.Lon, JamesAM, Jeronimo, Kvdveer, Lazman, Lemuel Gulliver, Leonard Vertighel, Loupeter, MIKHEIL, MadGeographer, Markussep, Mediafreaks, Mircea cs, Myasuda, Nk, Olessi, P-A., Pjetter, Rangolmaarn, Rhymezz, Rjwilmsi, Rmhermen, Roland zh, SHARU(ja), Schwyz, Spitfire, Stemonitis, Svick, TGCP, Ultramandk, Zacharie Grossen, 9 anonymous edits

Pavia *Source*: http://en.wikipedia.org/w/index.php?title=Pavia *Contributors*: A2Kafir, Ajrocke, Americanego, Andrea moro, Andreas Kaganov, Arch dude, Ariasne, ArnoLagrange, Arthena, Attilios, Bahnfrend, Bill Thayer, CARAVAGGISTI, Caeruleancentaur, Calian, Chris the speller, ClementSeveillac, Cntras, D6, Daderot, Dalobuca, David Parker, Doug Coldwell, Emanuele Saiu, Euchiasmus, Everyking, Frokor, G.-M. Cupertino, Ggonnell, Ghirlandajo, Gogafax, GreatWhiteNortherner, GreggHouse, GregorB, Gryffindor, Gugu102, Gusterbuster, HeartofaDog, Hede2000, Hmains, Honbicot, Hu12, Hypertrek, Ian Spackman, J04n, JASpencer, Jeronimo, JinJian, Joseaperez, Jpbowen, Jujutacular, Kinu, Kribbeh, Kwamikagami, Leptictidium, LindsayH, Looxix, Lucifero4, LuisGracias, Lvcretivs, MJCdetroit, Macrakis, MadGeographer, Marco.bozzi, Marek69, Matteo81, MauroVan, Mel Etitis, Mindmatrix, Mirgheca, Miwanya, Monegasque, Numbo3, OwenBlacker, Paginazero, Pearle, Plasticspork, Polylerus, Postlebury, Rich Farmbrough, Rio Cicica, Romanm, Semmelweiss, Sjc, Slawojarek, Srnec, Ta bu shi da yu, Taketa, TarmoK, TheCoffee, Theologiae, Thomo12, Tomi, Umedard, Unyoyega, Wetman, Wikigaz, William Avery, Wojtek 1956, X!, 93 anonymous edits

Italian_language *Source*: http://en.wikipedia.org/w/index.php?title=Italian_language *Contributors*: --, -Majestic-, -kayac71-, ..GaToR..KiNg.., 12jchristensen, 2help, 65.197.2.xxx, A-giau, ABF, ACW, ANGEL1312, Aaker, Aaron Einstein, Abbabash, Acromero12, AdRock, Adgpedia, AdjustShift, Adjwilli, Aelwyn, Aergniop, Aeusoes1, Afc0703, AgentPeppermint, Agoston, Aidan Elliott-McCrea, AimLook, Aitias, Ajox, Alansohn, Albeiror24, Albertopelis, Alejo2083, Alessandro.pasi, Alex '05, AlexR.L., AlexanderWinston, Alexandre8, Alfio, Alonsoavellaneda, AlphaEta, Altenmann, Aly95, Amire80, And we drown, Anders Feder, Andre Engels, Andres, Andrew Lancaster, Andrwsc, Andy Marchbanks, Anetode, Angelo Somaschini, Angelo.romano, Angiolo77, Angr, Angusmclellan, Antandrus, Antonio Lopez, Antun Gustav, Anypodetos, Arael2, Aranel, Arbitrary username, ArglebargleIV, Army1987, Ashdurbat, Ashenai, Asnac, Astroview120mm, Atif.t2, Atitarev, Atlant, Attilios, Aubadaurada, Auslli, Avala, Avicennasis, Aytharn, BGordon, BONGU SEAN, Bakabaka, BalticPat22, Baristarim, Bart133, Battlefield1918fan, Bcorr, Behemoth, Beland, Bemoeial, Ben Ben, Benwing, Bermuda-Russian lover556, Bestiarosa, Betterusername, Bfinn, Big water, Bigdaddy1204, BilCat, Bixxio, Blanchardb, Bletch, Blowski, BlueMars, Boffin, Bolivian Unicyclist, Bongwarrior, Booksworm, Boraczek, BorgQueen, Borock, Bowman, Brewcrewer, Brian0918, Brianga, Brion VIBBER, Broquaint, Bruinfan12, Brutaldeluxe, Bukkia, Burschik, Byrial, Byssinus, Byubatch, CARLMART, CJGB, CWenger, Calcwatch, Calmer Waters, Caltas, CalumH93, CambridgeBayWeather, Cameron Nedland, Can't sleep, clown will eat me, Capi crimm, Capricorn42, Captin Shmit, Carlobortone, Carlossuarez46, Carnby, Carre, Cassowary, Cast, Cedric Diggory, Chamdarae, Chameleon, CharlesMartel, Chaseparker, Che829, Cheese363, Cherylpok, ChiZeroOne, China543, ChongDae, Chowbok, Christopher Sundita, Chuckles220, Ciacchi, Cielomobile, Cioni69, Circeus, Ckatz, Clap, Cmonticc, Coemgenus, CommonsDelinker, Contributor777, Conversion script, Corti, Courcelles, Cratbro, CrazyChemGuy, Creedence, CristianoMacaluso, Croquant, Cryistyian, Culture69, Curb Chain, Curps, DCDuring, DO'Neil, DTOx, Daler, Damian Yerrick, DanMS, Danger, Daniel32708, Danny, Danski14, Danutz, Darwinek, Daviboz, Daviduzzu, Dbach, Dbachmann, Dcoetzee, DePiep, Deb, Delldot, Democracy to information, Demonkey36, Denisarona, Denisutku, Deor, Deviathan, Dgw, Diego, Dionix, Discospinster, Dissonancetheory, Dk1919 Franking, Doctorevil64, Don4of4, Donald Albury, Donarreiskoffer, Doubleplusjeff, Dougofborg, Drepanopulos, Drmaik, Drphilharmonic, Duoduoduo, Duomodimilano, Dwo, Dylan171, E.P.Y. Foundation, EdEColbert, Edward, Edward.oneill, Eequor, Ehrenkater, Eleassar, Electriceel, Elf-friend, Eliana619, Ellywa, Emrrans, Enok, Eob, Eranb, Ercole, Erianna, Esperant, EugeneZelenko, Exhuy, F l a n k e r, Fabio273, Faizaguo, Fang Aili, FastLizard4, Fawcett5, Felisopus, Femto, Fertuno, Fibonacci, Fiet Nam, FilipeS, Finngall, FlavrSavr, Flewis, Fobos92, Foros2000, Foxj, Frankenpuppy, Frecklefoot, Frerin, Friedfish, Fsolda, Fungusnunchuck, Funnyhat, Furrykef, Future Perfect at Sunrise, Fwb44, G.a.ys are di$gusting!, GJo, Gaius Cornelius, Garzo, GattoRandagio, Gbertoli3, Gdalle, Gdarin, GerardM, Gfoley4, Giacomo Volli, Gianfranco, Gibmetal77, Giftlite, Gilgamesh, Gilliam, Gioto, Gnhxwarri0r3, Gnowor, GoHawks4, Goochelaar, GraemeL, Grafen, Graham87, Grant65, Greman Knight, Grifter72, Grika, Gronky, Grossdomestic, Gspahr, Guido Gonzato, HQCentral, Hadal, Haeleth, Hak-kâ-ngìn, Hal8999, Haleth, HalfShadow, Hayden120, Hcholm, HeartofaDog, Helenarose1, Helios, Hemzlee, Henleydude, Heron, HexaChord, Hintha, Hippietrail, Hippophaë, Honza Záruba, IANVS, INyar, Ian Spackman, Iceager, Ihope127, Iketsi, Ikiroid, Ilario, Ilcioni, Imroy, Indnwkybrd, Insanity Incarnate, Inslide, Iridescent, Ironplay, Isabel100, Ispanyolcom, Italian boy, Italianoinspiritu, Ivirivi00, J. 'mach' wust, J. Finkelstein, J.delanoy, JB82, JDDunn9, JHunterJ, JYi, JaGa, JackLumber, Jahangard, JamesBWatson, Jay-Sebastos, Jbinder, Jcarroll, Jchthys, JdeJ, Jdhernan, Jerelott, Jfdwolff, Jiddisch, JimVC3, Jmabel, Jncraton, Joel B. Lewis, John, JohnCD, Joiede, Jojit fb, Jon Harald Søby, Jon Hawk, Jon cristaldi, JonHarder, Jonathan Grynspan, Jonkavelli, JorisvS, Jose77, Joseph Solis in Australia, Joy, Jsdixie uk, JulieADriver, Junesun, Jusdafax, K4zem, Kahuroa, Karmosin, Kaspo, Katalaveno, Katarighe, Kayac1971, Kegilbert, Kelly Martin, Kevlar67, Khoikhoi, Kinderjr, King of Hearts, Kingkobra123, Kjoonlee, Kman543210, Kmhkmh, Kneazle, KnowledgeOfSelf, Kntrabssi, Koavf, Kokiri, Kotabatubara, Kramden4700, Kromudh, Ks 7508, Kukini, Kuru, Kwamikagami, Kwertii, L Kensington, LAX, Lacrimosus, Laedan, Lambiam, Laura Lauretta, Lesnail, LewisDodgson, Lguipontes, LibLord, Lightmouse, Lilac Soul, Limbo socrates, Linda Martens, Lindpaik, Lindum, Lingua21, Livajo, LjL, Llayumri, Lornova, LovesMacs, Luckylick, Ludraman, Lugui65977, Luis wiki, Luka Jačov, Lupin, Lupo1982, M P M, M baptiste, MER-C, MJDTed, MKS, MadGeographer, Malhonen, Man vyi, Mani1, Marco, Marek69, Mariokempes, Mark Hyman, Markeilz, Marnen, Massimino, Matithyahu, Mattbr, Maxjurcev, Meaghan, MeekMark, Mega Prime, Mendelssohn, Mentifisto, Mgiganteus1, Mic, Michael Hardy, Michaelas10, Michaelschmatz, Mijzelf, Mikils, Milton Stanley, Mirandolese, Mircealucescu, Miskwito, Mithridates, Mjasfca, Modernist, Monteitho, Montrealais, Monty845, Moretti80, Mr.Right.Perfect, Munci, Mwanner, Mzajac, N3mesis, NHRHS2010, Naddy, Naelphin, Nalvage, NavalNut23, Nebbione, Ned Scott, Nehwyn, Nemo bis, Neverquick, Nikkaplzferreal, Noctibus, Nohat, Norm mit, NorwegianBlue, NotAnonymous0, Novaguy1968, Nri06, Nsaa, Numbo3, Oda Mari, Ohmscott, Ohnoitsjamie, Oleander122, Oliver Mundy, Oliverbeatson, Olivier, Ondewelle, Opinoso, Orodeo, Ortonmc, Orzetto, Osman93, OwenBlacker, Owenmadison, Oxymoron83, P-A., P36ad, Pacifista, Pan Brerus, Panairjdde, Paper1423, Pascar, PasswordUsername, Paul G, Paul-L, Paul0559, Paulbrock, Pazzocane, Peacock28, Perique des Palottes, PersOnLine, Peter Isotalo, Ph7five, PhilKnight, Philip Stevens, Philip Trueman, PhilipC, Phillip J, Philx, PhnomPencil, Pifactorial, Pippu d'Angelo, Pne, Poccil, Prodebugger, ProhibitOnions, Proofreader77, Purpleturple, Pzoxicuvybtnrm, Qqqqqq, Quiensabe, Quuxplusone, R. fiend, RPlunk2853, RadiantRay, RafaAzevedo, Raffaele Megabyte, Randhirreddy, RandomP, Randujar620, Rbfafe, Reconsider the static, Redcorreces, Redgolpe, ResearchRave, Retaggio, RexNL, Reywas92, Rich Farmbrough, Rivertorch, RoadTrain, Roberto Mura, Rodhullandemu, Rofl, Romanm, Ronline, Rothorpe, RoyBoy, Rrburke, Rsrikanth05, Rui Gabriel Correia, Russia77798, SEWilco, SPQRobin, SWAdair, SaberBlaze, Saintamh, Sakisg088, Salamurai, Samstayton, Sanmartin, Sbo, Sburke, Sceptre, Sebasbronzini, Sebesta, SemperBlotto, Sfo62, Shakawakasera, ShawnHath, Shirt58, Shreevatsa, Sicilgolem, Sicilianmandolin, Sirmylesnagopaleentheda, Sknute, SkyMachine, Sluzzelin, Smalljim, Smith120bh, Snowolf, SoCalSuperEagle, Soap, Sofa jazz man, Spacebirdy, Spearhead, Spellmaster, Spiff666, Splat, Squids and Chips, Squodge, Srnec, Staeiou, Starfight9600, Stephan Schneider, Stephen C. Carlson, Steverapaport, Stevvvv4444, Str1977, Sumail, Supasaru, Supertouch, Surcouf, SuzanneIAM, Svampen, Sólyomszem, TBH, TaliSaar, Tamfang, Tano-kun, TarisWerewolf, TastyPoutine, Taulant23, Taxman, Tb20000, Tb200000, Tgeairn, The Bushranger, The Show-Ender, The Thing That Should Not Be, The.Digital.Bedouin, TheMexican2007, TheOtherJesse, Theanthrope, Thecheesykid, Thijhgf, TicketMan, Tiddly Tom, Tide rolls, TimBentley, Time3000, Timwi, Tintenfischlein, Tjmayerinsf, Tnxman307, Tobias Conradi, Tolken, Tomi, Tomy B 2000, Tono-bungay, Torero, Townlake, Tpbradbury, Trc, Trompeta, Tropylium, Tudap84, Twaz, Uky123, UtherSRG, Vadac, Vaniba12, Vashtihorvat, Vasiliy Faronov, Velella, Ventifax, Vercillo, Vianello, Viator slovenicus, Vina, VivaLeet, Vrenator, Ward3001, Wavelength, Werdan7, Weregerbil, Wertuose, Wetman, WhisperToMe, Wiki Wikardo, Wiki alf, WikiLaurent, Wikiacc, Wikianon, Wikipedius, Wizzard2k, WolfgangFaber, Woohookitty, Worldlypeeps, XGustaX, Xezbeth, Xtreambar, Yanksox, Yeclense, YellowMonkey, Yes0song, Yom, Yupik, Zach4636, Zachlipton, Zaheen, Zambia95, Zanefletcher5, ZeneizeForesto, Zocky, Zoe, Zosch, Zyxw, Zyztem2000, Zzuuzz, سای, فشا, عقیل, . 1900 anonymous edits

Venice *Source*: http://en.wikipedia.org/w/index.php?title=Venice *Contributors*: *drew, -Midorihana-, 0lorenzo0, 151.24.144.xxx, 19.168, 2D, 2help, 3dnatureguy, 5 albert square, A bit iffy, A.K.A.47, ABF, Acheron, Adam Bishop, Adamacious, Adamrock, Adesterre, Adityam001, AdjustShift, Aecis, Aitias, Aivazovsky, Ajaynator, Ajdz, Akumy, Al-94, Alan Peakall, Alangdon86, Alanscottwalker, Alansohn, Alchemist Jack, Aleclod, Alexander Domanda, Alison, Allen3, Allens, Alluini, Alsandro, Alvisedm, AlwaysOnion, Alxndr, Amandajm, Amazonien, Amirtahataebi, Amoffit, Amuser2007, Andre Engels, Andrea biffi, Andrea.machiavelli, AndreaFox2, AndreasJS, Andres, AndrewHowse, Andrewmc, Andy M. Wang, Angelo De La Paz, Animum, Ankit Maity, Anna Lincoln, Anonimu, Anonymous Dissident, Antandrus, Aphaia, Aphasia83, Apparition11, Aranel, Areed145, Argyll Lassie, Arjayay, ArnoLagrange, Arpingstone, Arru, Arthena, Artjt, Aschenbach111, Attilios, Avicennasis, Avoided, Avono, AznBurger, B. Nuhanen, BD2412, Bachrach44, Badanedwa, Bantman, Bassbonerocks, Bastique, Bcnviajero, BeIsKr, Beetstra, Before My Ken, Bento00, Berserker27, Betterusername, Bfx12a9, Bhgriffdawg, Big Brother 1984, Bigtimepeace, Bigturtle, Bility, Bill Thayer, Binabik80, Binksternet, Binthemix, Biruitorul, Blahaccountblah, Blanchardb, Blisco, Blreyes, Bluemask, Blufive, Bobblewik, Bobo192, Bomac, Bonadea, Bongwarrior, Bookgrrl, Boricua1023, Bqwriter, Brad78, Bradyculous, Bragadin, Bramfab, Branddobbe, Briana676, Brutaldeluxe, Bsadowski1, Buaidh, Bxmouse94, CALR, CARAVAGGISTI, CDN99, CRKingston, CWii, Cabe6403, Cactus.man, Caerwine, Caltas, Calu2000, Can't sleep, clown will eat me, CanadianLinuxUser, CapPixel, CapitalR, Capricorn42, Captain-tucker, Caranicola, CardinalDan, Carlon, Carnby, Caroline Sanford, Casper Claiborne, Cataclysm, Catgut, Cburnett, Cdfitch1, CeeGee, Celiacozens, Century0, Cerobb01, Cferrero, Cgialloreto, Cgingold, Cglennon, Chamblyen, Chance Harper, CharlesMartel, Charvex, Chaser, Ched Davis, Chelsealb, ChimpanzeeUK, Chin frwnd, Cholmes75, Chris 73, Chris the speller, Chrisemms, Chrislk02, Chrism, Christopher Parham, Chumchum14, Chuunen Baka, CiTrusD, Cikraji, Citytransport.info, Claubelt, Cleror, ClockworkSoul, Closedmouth, Clossius, Cmdrbond, Cnchina, Cody12343, Coffeezombie, Colonies Chris, Colors99, Colt .55, Comfychaos, CommonsDelinker, Computerjoe, ConCompS, ConanBaltar, Condem, Confused monk, Conjohn, Conkle, Connected2daworld, Conte Giacomo, Conte di Cavour, Contributor175, Conudrum, Conversion script, Corlyon, Corndog99, Cortomaltesevenezia, Cosal, Cplakidas, Crazydaisy96, Cricobr, Crimson1984, Criticalthinker, Csabaengel, Cuchullain, Curps, Cyanolinguophile, Cymy, D6, DBigXray, DMacks, DTRY, DVD R W, Dadzone, Dajoker11, DandyDan2007, Danger, Daniel5127, Dannycas, Dante4848, DarkFalls, Darkwind, Darth Panda, Daubmir, Daufer, Dave1986, Davewild, Daviboz, David Kernow, David Parker, Davinci18, Dbachmann, Dbtfz, Ddama, DePiep, DeadEyeArrow, Deadfather, Dekaels, Delirium, Demiurge, Den fjättrade ankan, Denaya, Deprisa, Der Golem, DerBorg, DerHexer, Derek Ross, Dereksn, Dersh, Deville, Dfrg.msc, Digibrown, Dimboden, Dina, Dirkbb, Djanvk, Djheini, Djsasso, Dlae, Dna-webmaster, Docpi, Docu, Dogru144, Doisneau, Domino theory, DoubleBlue, Dougofborg, DougsTech, Dourios, Dpv, Dr. Zaret, Drphilharmonic, Dspradau, Dusselmann, Dyllholio, EH101, ESkog, Eastangerine, Eclecticology, Edtk, Edward, Egoffman, Ehamberg, El C, El Greco, Eleassar, Elekhh, Elektrik blue 82, Elibaba14, Elidauntheleve15wizard, Elijah scullin, Eliz81, Ellymental, Elonka, Eminem8867, Emustonen, Enchanter, Engunneer, Enochlau, Enti342, Enviroboy, Epbr123, Erauch, Erianna, Ettrig, Eugenio Archontopoulos, Euratlas, European Citizen, Evil Monkey, Evlekis, Excirial, Extendedtraveller, Extransit, F davey jones, Fab, Fabrictramp, Factlord, Fatasslol007, FayssalF, Fenners, Ferkelparade, Fertuno, Fieldday-sunday, Filiocht, Finlay McWalter, Fireworks, Fish1941, Floria L, FlyingToaster, Forkwiki, Foscari, Fotogian, FourTildes, Foxj, FrancescoMazzucotelli, Francs2000, Frankie54, Frankie773, Fred Lane, Frehley, FreplySpang, FreshBulletTime, Fritzpoll, Froid, Fsu5, Fuhghettaboutit, Fyreboy34, Fæ, Gabbe, Gaijin Ninja, Gaius Cornelius, Galati, Galoubet, Gamereo, Garrytowns, Gatty1, Gazimoff, Gcontini, General Wesc, Georgiasouthernlynn, Ghepeu, Ghouston, Giacomo59, Gianfranco, Giano, Gilgamesh, Gilliam, Gininho, Giov9, Giovanna123, Giraffedata, Giuseppe Danieli, Glane23, Glenlarson, Godmar02, Gogo Dodo, GoingBatty, Golbez, Goochelaar, Googlebooplehead123, GordonGraff, Goskooter, Goumbaguy, GraemeL, GraemeLeggett, Grafen, Graham87, GrahamBould, Green Giant, GregLoutsenko, GregorB, Gregorik, Grr, Grunners, Gryffindor, Gugganij, Gugilymugily, GumTree, Gump Stump, Gurch, Gurchzilla, Gvf, Gwernol, Gwj2001us, Gxabier, Gz33, Gzuckier, Gökhan, Haakon, Hadal, Hahnchen, Hamdan Al Darei, Hamiltondaniel, Hans-Jürgen Hübner, Harp, Havelock22, Hayden120, Haza-w, Hdt83, HenryLi, Heron, Herschelkrustofsky, HexaChord, Heydude88, Hfastedge, Highcheifpetermavia, Hismattness, History Packet, Hmains, Hobartimus, Hohenloh, Hon-3s-T, Honbicot, Hotcrocodile, Howcheng, Huntster, Husond, Hydriotaphia, Hydrogen Iodide, Hzenilc, IBook of the Revolution, ICarlz, Ianb, Idon'texist, Igiffin, Immunize, Indon, Infocidal, Infrogmation, Insanity Incarnate, Ipaat, Iridescence, Iridescent, Irishguy, Irtiqaa, Islander, It.wiki:Twilight, Itafran, Itai, Italfan, Italianstud-lina2, Ivana1, Ivirivi00, J Milburn, J.delanoy, J04n, JAF1970, JASpencer, JB82, JDX, JFLYONS, JForget, JJGeneral1, JNW, JaGa, Jacek Kendysz, JackofOz, JackyR, Jake Wartenberg, Jalen, James086, JanCeuleers, Javits2000, Jayzara, Jbrener, Jcr2, JeLuF, Jeanne boleyn, JediActor1998, Jeepday, Jeff G., Jeffrey Mall, Jeppiz, Jerome Charles Potts, Jeronimo, Jevansen, Jezhotwells, Jhaverkamp, Jheald, Jiddisch, Jim Bell, JimVC3, JinJian, Jivecat, Jjasi, Jmlk17, Jncraton, Jobeard, JoeSmack, Joergsam, Joevsimp, Joey80, Jogers, John K, JohnCD, JohnMorra, JohnnyB256, Jonjoes express, Joowwww, Jordz, Jorge Stolfi, Josiah Rowe, Joy, Jpv210, Jrdioko, Juanpdp, Jusdafax, Jusjih, Jvian, Jweiss11, Jyril,

K.lee, K1Bond007, KF, KFP, KJS77, Kaare, Kaaveh Ahangar, Kafka Liz, Kahastok, Kaiser matias, Kallath, Kaphkian, Karthics4, Karvan g, Kellerkind, Kelvinc, Ken Gallager, Kenneth02, Khvalamde, Kingpin13, Kipala, Kirill Borisenko, KnowledgeOfSelf, Kntrabssi, Konstantinkoll, Kosebamse, Kpjas, Kristenryee, Kross, Krowms, Ksizova, Ksy92003, Kungfuadam, Kurt S Koller, Kuru, Kwamikagami, L Kensington, LFaraone, Laenfant, Lammidhania, Leandros, LeaveSleaves, Leilania, Lelavr, Lemuel Gulliver, Leolohwj, Levineps, Lfh, LibLord, Lidos, Lightmouse, Lilac Soul, Limideen, Lithpiperpilot, Little Mountain 5, LittleJewbaca, Llort, Londonlnd, Lorenzo ciao, LorenzoB, Lori89, Lotje, Louie Franco, Love 2 B Fair, Lradrama, Lrgs11855, Lrgs11889, Luc19, Lucaf1, Luckas Blade, Luctin, Luigi 28, Luke hardy, Lukobe, Luna Santin, Lupin, Lynchboy, Lyssachelle, M.tieuli, M4t10, MER-C, MJCdetroit, MZMcBride, Mabalu, Maelnuneb, Magister Mathematicae, Magnus Manske, Magog the Ogre, Maha ts, Man vyi, Manfroze, Mannafredo, Manomow, Maralia, Marce1979, Marcok, Marek69, Mareklug, Mario1010, Marj Tiefert, MarkMarek, Martin Garrett, Masonbarge, Massimo Catarinella, Masterjamie, Matej1234, Materialscientist, Mathpianist93, Mattgirling, Matthew Yeager, Mattytay, Mauls, Mav, Maximus Rex, Maxis ftw, Maymay 1699, MeekSaffron, Melamed katz, Mentifisto, Metb82, Metropolitan90, Metstotop333, Mfiskum, Mhoenig, Michael Hardy, MichaelBillington, MichaelTinkler, Michellecrisp, Midnightdreary, Mierlo, Mike D 26, Mike R, Mike Rosoft, Mikhailov Kusserow, Mimmo46, Mindmatrix, Mintleaf, Miqueli, MisterHand, Misza13, MithrandirAgain, Mjkelton, Mkoval, Mksmith, Modster, MollyTheCat, Monkey Bounce, Montrealais, Moonraker, Moquist, Moralis, MorganaFiolett, Moshe Constantine Hassan Al-Silverburg, Mr Adequate, Mr. Wheely Guy, MrMime, MrTarry, Mrdice, Mrmaroon25, Mrschimpf, Mrsly, Mtaylor848, Mukkudal, Mulad, Murali mohan1, Mushroom, Musica e Gatti, Myanw, MykReeve, Mym-uk, NMChico24, NYKevin, Nabla, Nakon, Nalta, Narcis010, Narrasawa, Nasnema, Nattfodd, Nauticashades, NawlinWiki, Necrothesp, Nedrutland, NeilN, Neizod, NellieBly, Netalarm, Neurolysis, Neutrality, Never Mystic, Neverquick, Nevilley, Newton2, Niceguyedc, Nick Ottery, Nick mccallum666, Nightscream, Nihiltres, Nivix, Nkatz22, Nlu, Noclador, Noctibus, Noluogo, Non-dropframe, Norm mit, NorwegianBlue, Nova77, Nowiki123, Nsaa, Nskillen, NuclearWarfare, Nuttycoconut, Nv8200p, Nymble, Oberiko, Ocee, Oda Mari, Odedee, Odie5533, OettingerCroat, Ohconfucius, Ohms law, Ohnoitsjamie, Old Moonraker, Oldwindybear, Olivier, Ollee, Ombudswiki, Omicronpersei8, Onejenu, Only, Onorem, Oobopshark, Openproject75, Optimus82, OrangeDog, Oreo Priest, Other-thing, OverlordQ, Oxymoron83, Ozanpol, P. S. Burton, P.K.Niyogi, P199, PDXblazers, PL290, Paginazero, Palazzo, Panairjdde, Panarjedde, Paramore1357, Paste, Pathoschild, Patrick, Patstuart, Paul-L, PaulHardwick, Paulac619, Peaked, Pebbles1979, Pedroshin, Pekaje, Pelican39, Pepper, Permethius, Persian Poet Gal, Peruvianllama, Peter Isotalo, Peyre, Pharaoh of the Wizards, Phenylalanine, Phgao, Philaweb, Philip Trueman, Phillip J, Piano non troppo, Pianoplonkers, Piccolo Modificatore Laborioso, Pichpich, Pigsonthewing, Pilotguy, Pinethicket, Pinkfinkle, Piotrus, Plasticspork, Pleclech, Pmsyyz, Pollux95, Ponyo, Popadopolis, Porphy, Porqin, Possum, Power piglet, Profe$$or, Pronuncia, ProveIt, Psb777, Pstanton, Pubaquoc, Puchiko, Purplebackpack89, QEDquid, Qazwsx13, Quadell, Quaeler, Quasipalm, Queenofthewilis, Quincy, Qwavel, R'n'B, RA0808, RHaworth, RJHall, RReis, Raidon Kane, RainbowOfLight, Random86, RandomP, Ray Chason, Raymondwinn, Razorflame, Rdsmith4, RedWolf, Redeagle688, Redthoreau, Reedy, RegentsPark, Regioneveneto, Remoq, Renaissancee, Renata, Renata3, RexNL, Reywas92, Rfl, Rgoodermote, Rich Farmbrough, Rich257, Richard Keatinge, Richard75, Rjensen, Rjwilmsi, RobertBMeadows, RobertG, Robertsteadman, RodC, Rogitor, Rohan Jayasekera, Ronhjones, Ronz, Rosarinagazo, Royal Blue Jersey, Royalguard11, Roybb95, Rrburke, Rror, Rscano, Rturus, Russavia, Russian Luxembourger, RyanCross, S h i v a (Visnu), SECProto, SEWilco, SF88, SJ Morg, SJP, Saffron Blaze, Saga City, Sahmeditor, Salvo46, Savidan, Scanos, Scotthatton, Scoutersig, Sdoerr, Sdw25, Seamusmckenna, Seanlin101, Seb az86556, Segv11, Selket, Selma Kaufmann, Semolo75, Sergiojbneves, Sganjam, Sha-256, Shadeofmyheart, Shanes, Shantavira, Shawn in Montreal, ShelfSkewed, Shell Kinney, Sherlock, Shesalt, Shimgray, Shoessss, Shortsmicsnigore, Shultz, Sicilianmandolin, Siegele Roland, SilkTork, SimonP, Simoncursitor, Sintaku, Sintonak.X, SiobhanHansa, Sir Nicholas de Mimsy-Porpington, Sjakkalle, Sjc, SkyMachine, Skyguy414, Sladen, Sluzzelin, Smably, Smashville, Smb1001, Smoth 007, SmthManly, SniperduckW, Snowolf, Solarat, Solipsist, Solitude, Some jerk on the Internet, Sos10, Southernreference, Speer320, Spenmac, Srnec, Stan Shebs, StaticGull, Steevm, Stefano Remo, Stephenb, Steven Zhang, Stevenmitchell, Stickee, Stubblyhead, Stwalkerster, Sunray, Super Mario, SuperSmashBros.Brawl777, Superlux74, Sverdrup, Svetovid, Sweeney0908, Syferus, SyntaxError55, THEN WHO WAS PHONE?, TJFadness, Tango7174, Tapiros, Techhead7890, Teiladnam, Terence, Terminal157, Test123abc321, Texterone, Tfinelli, The Cool Kat, The Rambling Man, The Thing That Should Not Be, The undertow, TheDoctor10, TheKMan, TheLongTone, TheTrojanHought, Thehelpfulone, Theloniouszen, Theme97, TheoClarke, Theologiae, Thepatriots, Theroadislong, Thewayforward, Thingg, ThinkBlue, Thome66, Thyrsis, Tide rolls, Tiger888, TigerShark, Tigermike15, TimBentley, TimDuncan, TimHalldor, Timeineurope, Tinatin8, Tkreuz, Tobby72, Tobias Conradi, Tocino, Tom Gore, Tommy2010, Tontotti, Tony1, Topsydog321, Torrazzo, Tourtux9, Townlake, Tpbradbury, Trasman, Travelbird, Travellingbazaar, Treemother199, Trident13, Trimalchio, TruckCard, TrulyBlue, Trystero, Tshamilton, Tsja, Tuminof, Turbokoala, Turnstep, TurtleMelody, Undangerman, UniverseKing, Unofeld781, Urbanfurniture, Urgos, User2004, Utcursch, VX, Valentinian, Vanished User 1004, Vanished user 39948282, Vegaswikian, Venetian, Venicetourismo, Versus22, Vicenarian, Villefort1790, Viva-Verdi, Vorchard, Vrenator, W1k13rh3nry, WTGDMan1986, WWGB, Wackytaco1234, Waggers, Wai Hong, Wallie, Walton One, Warofdreams, Wavelength, Wayne Slam, Wcsdep2313, Wechytee, Weenieroast, Wetman, WhisperToMe, Whytecypress, WickEd Z, WieDiRoc, Wiki alf, Wiki13, WikiLaurent, Wikimeow, Wikisabo, William Avery, Wimt, Wingfield-digby, Wk muriithi, Wonder al, Woohookitty, Wtmitchell, XJamRastafire, Xaliqen, Xanzzibar, Xiaoyu of Yuxi, Xn4, Xnuala, Xtreambar, Xxanthippe, Y2kcrazyjoker4, Yahia.barie, YellowMonkey, Yeonhum1357, YeshuaDavid, Ygoloxelfer, Ynysgrif, Yodaman35633, Youandme, Your a dummy, Yporat, Yuje, Yurivict, Z0nkandzig, Zabop1, ZacBowling, ZacharyS, Zachlipton, Zadcat, Zebra-police, Zer0faults, Zerblatt, Zidonuke, Zomic13, Zondor, ZooFari, Zoranos, Zouf, Zscout370, Zserango, Zzuuzz, Δ, Владимир Шелянин, , პაატა შ, 3003 anonymous edits

Switzerland *Source*: http://en.wikipedia.org/w/index.php?title=Switzerland *Contributors*: (jarbarf), -- April, 05, 100110100, 130.60.153.xxx, 159753, 16@r, 334a, 5dsddddd, 661kts, 7anonymous7, 7tests, 80sfaan, A-giau, A3RO, AB-me, ABF, AFA, Aaalroex, Aaker, Aaron Einstein, Aaronbrick, Abc518, Abrech, AbsoluteZero280, Academic38, Acalamari, Access Denied, Acelor, Achowat, Achtzehn, Acs4b, Adam Bishop, Adam78, Adamahill, Adashiel, Adhall, Adi4094, Aetheling, Aff123a, Agapius1987, Aguasde13013, Ahoerstemeier, Ahuskay, Airplane yow, Aitias, AkAiBaRa, Akanemoto, Akendall, Alan16, Alansohn, Aleph4, Alethiareg, Alex S, Alex.muller, Alex45, AlexTheGrand, Alexander Domanda, AlexiusHoratius, Ali K, Alivath, Alperen, Alphabravotango, Alphasinus, AltWiki, Altermike, Ambg, American1991, Amerika, Amorim Parga, Andajew, Andre Engels, Andrew Levine, Andrew33base, Andrewlp1991, Andrewrp, Android Mouse, Andy M. Wang, Andy Marchbanks, Andyindia, Angel,Isaac, AngelOfSadness, Anger22, Angr, Angusmclellan, Animum, Ann Stouter, Anna Lincoln, AnnaFrance, Anonymous Dissident, Anonymous anonymous, Answar, Antandrus, Antonio Gonzales, AntonioMartin, Anxa, Aoibheannniamh, Apiarius Flavius, Arad, Aranherunar, Armando, Arminescher, Army1987, Arne List, Arwel Parry, Ascend, Atitarev, Atumbabod, AuntFlo, Austin512, Austinthebookworm, Avala, Avaya1, Avenged Eightfold, AvicAWB, Avicennasis, Awpc, AxG, Axeman89, Ayrton Prost, Azazello, B. Phillips, BAAKO, BBT2005, BRG, BTLizard, Bababoef, Babybaby78, Backslash Forwardslash, Badenoch, Bakerccm, BalkanFever, Bambuway, Bananaclaw, Banpei, Barek, Basawala, Baseball Watcher, Baseballbaker23, Bazonka, Bcnviajero, Bdbfly1, Beano, Behemoth, Beland, Bellahdoll, Bender235, Benjiwolf, Benkok101, Bennylin, Benyon3, Bevo74, Bggoldie, Bhumiya, Bickeringwife, Bidabadi, Bigfujitsu, Bility, Bilko yes, BillC, Binary TSO, Bjarki S, Blake alan16, Blanchardb, Blavertu, BlazeJag, Bleach3, Blessthishouse, Bloody resident, Blue520, BlueMars, BlueNovember, BlueXen, Bluee Mountain, Bluerasberry, Bmoq, BoLingua, Bob247, BobIV, Bobblewik, Bobbo, Bobg123, Bobo192, Bodnotbod, Bogey97, Bokpasa, Bolo1729, Bongwarrior, Booksworm, Boost3000, BorgQueen, Borisblue, Bosonic dressing, Bourquie, Bradoni, Brainboy109, Brandon, Branlon, BrendelSignature, BrianGV, BrigadierGeneralBaptisms, Brighterorange, Brion VIBBER, Brisvegas, BritishWatcher, Broha8, Brrryan, Brunton, Brutaldeluxe, Bsadowski1, Buaidh, Bucephalus, Bud467, Bugnon, BuickCenturyDriver, Bumm13, Bunchofgrapes, Bunny-chan, C mon, C0nanPayne, C28martin, CIreland, CO, CONNOR-WALSH-95, CRE is king, CSZero, Cae prince, Caffein, CalJW, Calabe1992, Calabraxthis, Calbaer, Callum389, Calmargulis, Caltas, Cam147, CambridgeBayWeather, Camilo Sanchez, Can't sleep, clown will eat me, CanadianLinuxUser, Canderra, Canley, Cantanameras, Canterbury Tail, Cantus, CapitalR, CapnPrep, Caponer, Capricorn42, Captain Spleen, CardinalDan, Carlosforonda, CaseyPenk, Casper2k3, Cate, Catfish Jim and the soapdish, Catfishy456, Caumasee, Causa sui, Cbowers@kumc.edu, CdaMVvWgS, Cdc, CeeWhy2, Cenarium, Centrx, Cepal, Chanakal, Charvex, Chatfecter, Chiacomo, Chipmunkdavis, Chochopk, Chris Rocen, Chris the speller, Chrisch, Chrisieboy, Christianbrewerrox, Christopher Sundita, Chrono1084, Chronodm, CiTrusD, CieloEstrellado, Cinephile, City of Destruction, Cjrs 79, ClamIAm, Claude girardin, Cleduc, Clorox, Closedmouth, Clpda, Coffee, Cogito ergo sumo, Cokepepsi, Colonies Chris, Colt .55, Coltfan009, Commander Keane, Commando502, Common appeal, CommonsDelinker, Conglacius, Connormah, Conorbrady.ie, Contributor777, Conversion script, Coolgirl1934, Coolhawks88, Coppernut, Cordless Larry, Corti, Corvus13, Costesseyboy, Cptnono, Crazyjesenia, Crazywaters, Credema, Cremepuff222, Creol, Crhottie01, Cristian Cappiello, Cruzian, Cryptic, Cs-wolves, Cs219cheater, Csabi911, Css, Cub68134, Cunado19, Cuthbertgrant, CyberPanda, Cybercobra, Cyde, Cyhawk, CyrilleDunant, D, D4RK-L3G10N, D6, DG, DJ Clayworth, DJSoundBored, DMacks, DW, DaDoc540, Daa89563, Dacrazysmoose, Dahveed323, Dale-swiss, Dalstadt, Dancter, Dandabomb3, Danger, Danno uk, Danny, Dapete, Dardan Krasniqi, Darrendeng, Darthdj31, Davewild, David Gerard, David Kernow, David Liuzzo, David.Monniaux, DavidHolden, DavidNw, Davidfreund, Dawright12, Dayubcpd, Dbachmann, Dbenbenn, Dbfirs, De728631, DeadEyeArrow, Deadstar, Deathphoenix, Dedee73, Deh, Dekisugi, Delpino, Deltabeignet, Deltasigma, Demmy, Denelson83, Deor, Der Eberswalder, DerHexer, Dexter prog, Dgw, DiENass, DiamondShade, Dianaruttman1, DieZurcher, Diftong, Dimitrii, Diothoir, Directorstratton, Discospinster, Dittaeva, DivineAlpha, Dmn, Doc Quintana, Doc glasgow, DocWatson42, Doctorevil64, Docu, Dodo von den Bergen, Dogfish Jim and the Dixoap, Don Durandal, Donkeydick11, Dookymonkey, Dorftrottel, Dougofborg, Dr. Blofeld, DrKiernan, DrYak, Draconion devil, Dragonlord kfb, Drbug, Dreadstar, Dream Focus, Drepanopulos, Drew R. Smith, DriveMySol, Drmies, DrunkenSmurf, Dschwen, Dspradau, Dunkedun, Dust Filter, Dwo, Dwunderli, Dysepsion, E Pluribus Anthony, E Pluribus Anthony redux, E2eamon, ECube Global, EJF, ERcheck, ESkog, EZ, Eaefremov, EamonnPKeane, Eclecticology, EdBever, Ediacara, Edificia, EeepEeep, Efghij, Eggman64, Egmontaz, Ejk81, Ejs-80, El C, El Suizo, El estremeñu, El xeneixe12, ElPeste, Electionworld, Elockid, Eloquence, Emet truth, Emil Kastberg, Emmisa, EmpireKing, Enchanter, EncycloPetey, Energyfreezer, EnglishDude, Enigmocracy, EoGuy, Epbr123, Epicadam, Equendil, Erianna, Eric-Wester, Ericd, Ericoides, Esrever, Esteffect, Estel, Eu.stefan, Euratlas, EurekaLott, Europeanul, Everyking, ExRat, Eztoremember, FF2010, FJM, FJPB, Faeyd, Falcon8765, Fan-1967, FatherofAssBurgler, Favonian, Fayenatic london, Fdorville, Fede72, Felyza, Ferhengvan, Ferix, Ferkelparade, Filing Flunky, Filipdr, Filll, Fish-guts, Flatterworld, FlavrSavr, Flewis, FlieGerFaUstMe262, Floul1, Flowerpotman, Fluri, Fonzy, Foreigner, Forest Elf, Forumyoungblood, Fotaun, FotoPhest, Foxj, Fr34k, Fram, France3470, FrancoGG, Francs2000, Frasor, Fred Condo, Fredgoat, Freeminder, FreplySpang, Frichmon, Friends147, Friginator, From-cary, Frukko, Fsotrain09, Fudoreaper, Fungusbrain, Funnyhat, Furrykef, Future Perfect at Sunrise, Fuzheado, Fæ, G.A.S, G00dfella613, G026r, GC, GCarty, GL, Gabrieldawiki, Gadfium, Gail, Gaius Octavius Princeps, Gaj777, Gary King, Gayjin, Gazpacho, Gdo01, Geni, Geoff.powers, Georgegeorgegeorge, Geraldo Perez, Gerrit, Gestumblindi, Gggh, GiZiBoNG, Giftlite, Gilliam, Gingerskyler, Gioto, Girolamo Savonarola, Glane23, Glaurung, Glenn, Glenndsymons, Gman124, GoetheFromm, Gogo Dodo, Golbez, Goldom, Golgofrinchian, Good Olfactory, Goodlucca, Graeme Bartlett, Grafen, Graham87, GrandSapin, GrandpaDoc, Granpuff, Grayshi, Greatestrowerever, Green Giant, Green caterpillar, Gregory Shantz, Grillo, Grimey109, Gringo300, Groovyfairyfloss, Grosscha, Grundig, Gryffindor, Gsarwa, Gscshoyru, Guanaco, Guliolopez, Gurch, Guy Peters, Guywithmako, Gwernol, Gwguffey, Gyrothepyro, Gzornenplatz, Gökhan, Gürbetaler, HAMM, HJ Mitchell, Habib87, Hadal, Hairy Dude, HalfShadow, Hanacy, HangingCurve, Hannes Röst, Hans Dunkelberg, Harej, Harel Newman, Harp, HarrisMC, Harry the Dirty Dog, Harshapaulraj, Hashem sfarim, Hateless, Hayden120, Hdstubbs, Headbomb, Hemnarayan, Henrygb, Hephaestos, HexaChord, Hgilbert, Hiddekel, Highvale, Hiimtod, Historian19, Hmains, Hmusseau, Hoary, Hooriaj, Horgen, Hotcrocodile, House.au, Howdydoo, Hu12, HumanrightsFR, Huntington, Hushpuckena, Husond, Hut 8.5, Huw Powell, Huwr, Hydrogen Iodide, Hyyttaa, ICE77, IIIraute, IRP, Ianb, Ice Cold Beer, Icemuon, IchbinIch, Idaltu, Idinic, Ief, Iem, Igo4U, Immunize, Improv, InShaneee, Indon, Infrogmation, Inner Earth, Inside switzerland magazine, Insideswitzerland, Instinct, Intangible, Into The Fray, Ioannes Tzimiskes, Iridescent, IronGargoyle, IstvanWolf, Italiano111, ItsZippy, Itsmine, Iunaw, Ivirivi00, Ixfd64, Izzedine, J. 'mach' wust, J. Naven, J.Wright, J.delanoy, JForget, JFreeman, JNW, JVz, JYi, JaGa, Jab843, Jackalisaghost, Jackl, JackofOz, Jacobgibbons, Jacooks, Jaggsan, Jake Wartenberg, JakeVortex, Jakezing, Jakro64, Jam01, Jamcib, JamesBWatson, JamesR, JamesReyes, Jamiestahl, Jammann, Jao, Jaranda, Jauerback, Jayen466, Jayron32, Jbsongaku, Jcain1635, Jclemens, Jeanenawhitney, Jeff G., JeffDaniels, Jeremydillon, JerryFriedman, Jerryseinfeld, Jessejamesbisunia, Jezhotwells, Jhbuk, Jhendin, Jhjhjhfffffffffff, JiMidnite, Jiang, Jim10701, Jim1138, JimVC3, Jimspilman, Jj137, Jjhcap99, Jklumker, Jlmlax99, Jmaddux23, Joebob23, Joffeloff, John, John of Reading, JohnFromPinckney,

Johnandbob, Johnnycakes234, Jojhutton, Jon cristaldi, JonMoore, Joolz, Joost 99, Jordz, Jorunn, Jose77, Joseph Solis in Australia, JoshHolloway, José Ignacio Cabrejos, Jovianeye, Joy, Joysofpi, Jpeob, Jpowell, Jrielaecher, Jsl83, Juliancolton, Jumbuck, Jørgen, K10wnsta, Kabelleger, Kaf, Kaihsu, Kain Nihil, Kairos, Kaiserb, KaitlinC11, Kaitsu80, Kanatonian, Kane5187, Kangy, Kannie, Karl, KarlFrei, Karuna8, Kat90210, Katalaveno, Kbh3rd, Kbolino, Keenan Pepper, Keilana, Keimzelle, Kelisi, Kelnos, Kenbei, Kessler, Kevin Murray, Keycard, Khoikhoi, Killervogel5, Kim-Zhang-Hong, King Toadsworth, King of Hearts, Kingjezza, Kingpin13, Kintetsubuffalo, Kiss My Ass Bitch, Kitch, Klapper, Kman665, KnowledgeOfSelf, Kokiri, Kokpit, Kosovarvirg, Kotniski, Koyaanis Qatsi, Kozuch, Kr1st1deejay97, Krator, Krellis, Kseferovic, Ksnow, Kubigula, Kusma, Kwaku, Kwamikagami, Kylu, Kyphe, KyraVixen, Kzeynel, L Kensington, L95slovenia, LAAFan, LAX, LFaraone, LOL, La Pianista, LaMenta3, Lairor, Landon1980, Latka, Lazyfair, Lazylaces, Leandrod, LeaveSleaves, LeedsKing, Leflyman, Legendsword, LeoDV, LeoHeska, Leoni2, Leszek Jańczuk, Lethe, Levineps, Lezard, Lfgoette, Lfh, Lg1223, Liblamb, Liftarn, Lightmouse, Ligulem, Lijnema, Lilac Soul, Linburats, LineOfBlackCars, Lineface, Lithpiperpilot, Littrans, Lkaelin, Lkjhgfdsa, Lkmorlan, LlywelynII, Loadmaster, Lochness Monstah, Lollyz10, Longwayround, Lopokoko, Loren.wilton, LorenzoB, Lostcauses, Lostvalley, Lotje, Lprideux, Lsy098, Luca.gambazzi, Luffy487, Luna Santin, Lupin, Lupo, M0s6p, MBisanz, MER-C, MJCdetroit, MJD86, MKWat, MONGO, MPerel, MZMcBride, Macedonian, Mach2moo, MadGeographer, MadSurgeon, Madhero88, Maelnuneb, Magicalsaumy, Magore, Mahahahaneapneap, Male1979, MaliNorway, Mamawrites, Man vyi, Mandarax, Mangwanani, ManyFireflies, Maokart444, Marc Mongenet, Marco bisello, Marcworca, Marek69, Markb, Markonen, Marktreut, Marlinite, Marlonparker, Marnanel, Martin451, MartinUK, Martinwilke1980, Marzoregulus@hotmail.com, Mascia Luccova, Masterjamie, Matdrodes, Mateo SA, Materialscientist, Mathpianist93, Matt Crypto, MattieTK, Maurog, Max rspct, MaxSem, Maziotis, Mbalelo, Mboverload, McDogm, McSly, Mclay1, Mcmuffinpie, Mcmuffinz, Mdw0, Meaghan, Media anthro, Meegs, Meekywiki, Meertn, Megaman3020, Melaen, Melsaran, Mentifisto, Mentisock, Meow kitties, Merbabu, Mericasio, Merovingian, Mesgul82, MetsFan76, Mezigue, Mic, Michael Hardy, MichaelBillington, Michelle1234567890987654321, Mickey gfss2007, Midright, Midwestern Demigod, Mifter, Mikaey, Mikaveikka, Mike12123, Mike15564, Mikeo, Mileage, Millahnna, MindstormsKid, Minhnd, Minimac, Miquonranger03, Miraculousrandomness, Mirv, Mjgw, Mkilly, Mkztg, Mm.srb, Mmdoogie, Mo0, Moab1111111111, Momet, Momo erez, Monni1995, Monoecus, Montrealais, Monty845, Morana, Moreschi, Morio, Mossmen1531, Moumine, Mouse Nightshirt, Mr mormon, MrRadioGuy, Mrdeco, Mrholybrain, Mrxak, Msaini, Msikma, Muchosucko, Mufka, Mukitil, Mxcatania, Mxn, My76Strat, Mygerardromance, Myleslong, Mysdaao, Mzmujib, Mithrandir, N2e, N5iln, NByz, NCDane, NHRHS2010, NWill, Nach0king, Nagy, Naive cynic, Nakon, Nakos2208, Nathanael Bar-Aur L., Natl1, NawlinWiki, Nburden, Ncdoyle, Nellis, NeoChaosX, Nepenthes, NerdyNSK, Netkinetic, Netknowle, Neverquick, Nevilley, NewEnglandYankee, Ngagnebin, Niamh Smells, Nichalp, Nicholassharland, Nick, Nick Michael, Nickbenjohn1, Night of the Big Wind, Nightstallion, Nirvana2013, Nishant12, Nivix, NixonB, Njaelkies Lea, Njm0, Nkcs, Nlu, Nneonneo, Noah Salzman, Noisy, NonNobisSolum, Nonpawit, Nonplus, Nopetro, NorVegan, Nossac, Notredamefan747, Nova77, Nposs, Nsaa, Nthep, NuclearWarfare, Nudibranch2, Nuggetman, Numbo3, OL144, OOODDD, OSborn, Oda Mari, Odengatan, Off2riorob, Ohconfucius, Ohnoitsjamie, Ohwell32, Olivier, Omicronpersei8, Oneilius, Ordew, Oreo Priest, Osterluzei, Ottawa4ever, Ourai, OwenBlacker, Oxfordwang, Ozes, P4k, PAK Man, PBS-AWB, PFHLai, PL290, PRhyu, PZFUN, Padraig, Paine Ellsworth, Pajfarmor, ParisianBlade, Parmaestro, Party, Pascar, Pasu, Pathoschild, PatriceNeff, Patrick, Patrickmayor, Patstuart, Pattycakes13, Paul August, Paul Faller, Paulcmnt, Paulyeobee88, Pavel Vozenilek, Pax:Vobiscum, PeaceNT, Pearle, Pecksique, Pensionero, Pepper, Peter Horn, Peter.C, PeterEastern, PeterSymonds, Petersdick, Pethr, Petrb, Petropoxy (Lithoderm Proxy), Pg2114, Pgk, Pharaoh of the Wizards, Pharos04, PhilKnight, PhilMacD, Philip Trueman, Philippe317, Philpm930, Philthecow, Phoenix-forgotten, Pi, PiMaster3, Piano non troppo, Piccolo Modificatore Laborioso, Pigman, Piledhigheranddeeper, Pimplilboyahanicky, Pinethicket, Pippu d'Angelo, PizzaofDoom, Pizzarocks09, Pjetter, PlantHead, Plokijnu, Pne, Pokymon99, Pol098, Polaron, Postmortemjapan, Postoak, Pozcircuitboy, Pp.zg.hr, Praline97, PranksterTurtle, Prashanthns, Prince of History, PrincessofLlyr, Prodego, Proteus, Prylon, PsY.cHo, Psy guy, Ptah, the El Daoud, PurpleInk, Purpleturple, Pv2b, Pyrospirit, Qaka, Qasaqsuyu, Quadpus, Quercus basaseachicensis, Quiensabe, Quizimodo (usurped), Qwertasdf10, Qwyrxian, Qxz, R9tgokunks, RB972, RFBailey, RTC, Racerguy, RainbowOfLight, Raining girl, Rama, Ran, Random user 39849958, RandomAct, RandomOrca2, RandomStringOfCharacters, RaseaC, Ratemonth, Raven in Orbit, Raven4x4x, Ray Van De Walker, Raymondwinn, Razorflame, Rctycoplay, Rdsmith4, Rebug, Red Winged Duck, RedZebra, Redbombtoma, Redgolpe, Redhill54, Redmarkviolinist, Redsox11790, Redsxfenway, Regibox, Reinthal, Reisio, Rejectwater, Rejedef, Renaissancee, Renata, Revelian, Reverendgraham, Revolución, Revolutionary, RexNL, Rexlis, Reywas92, Riana, Rich Farmbrough, Rich257, Richard Arthur Norton (1958-), Richard New Forest, Richard.decal, Rick Block, RickK, Ricky81682, Ricky@36, Rickyiskandar, RiingoNYM, Riker wt, Rizalninoynapoleon, Rjwilmsi, Rl, Rlbarton, Rmhermen, Robert25, RobertG, Robertissimo, Rocket000, Rockymountains, Rodney Boyd, Roland Abt, Roland Kaufmann, Romanm, Ron Ritzman, Ronhjones, Ronin of Chaos, Rotovia, RoyBoy, Rredwell, Rror, Rsrikanth05, Rudolfhess456, Ruhrjung, Rumpelstiltskin223, RussBlau, Ryan McGuinness, RyanGerbil10, S charette, SDC, SMC89, SSZ, ST47, Saforrest, Saimdusan, Saippuakauppias, Salaciousd, Salion, Salvio giuliano, Sam Blacketer, Samantha555, Sameer.charles, Samrolken, Samuel.Mago, Samwb123, Sandstein, Sanghvimihir, Sango123, Sannse, Satori Son, ScaldingHotSoup, Schneelocke, Schutz, Scipius, Sciurinæ, Scohoust, Scott Gall, Scwlong, Sdnegel, SeanLegassick, Seaphoto, Search4Lancer, Seb az86556, Seba5618, Sebesta, Secfan, Secretlondon, Seki rs, Selket, Sellyme, Semmler, Sengstag, Seric2, Serte, Sewing, Shadow007, Shadowblade, Shadowslayer978, Shak-74M, Shamir1, Shandris, Sharkface217, Shawn in Montreal, Shawn2082, Shlomke, Shoessss, Shortstraw, Shredder2012, Shtop Shtealing from my Shtore, Sidonius, SigPig, Sikaolli, Silly rabbit, Silvermoonx, Silverxxx, Simcom, Simkiott, SimonP, Sims2aholic8, Sin-man, SineWave, Singularity, Sinn, Siqbal, Sir Richardson, Sjakkalle, Sk8er777, Skew-t, Skizzik, Skunkboy74, Skunkwaffle, SkyBlue eagle, Slowmover, SluggoOne, Sluzzelin, Slysplace, Smackdownvsraw, SmellyLilYou, Smirnius, SnowFire, Snowolf, Sockr44e, Soelke, Soman, Some jerk on the Internet, SomeHuman, Somnewise@yahoo.com, Soumyasch, Sourside21, South Bay, Southparkfan007, SpNeo, Sparr, Spellcast, Spettro9, SpookyMulder, Springnuts, Sputniktilt, Srrostum, Srushe, Staffwaterboy, Stalker X, Stan Shebs, Startstop123, Station1, Steel, Steinsky, Stemonitis, Stephan Leeds, Stephan Schulz, Stephenb, Stepheng3, Stevertigo, StradivariusTV, Strafpeloton2, Stuinzuri, Stupid Corn, Stw, Sue Rangell, Suitequai, Sullivanfam, Summer Song, Sunray, Supercoop, Supertask, Sureshot54, Suruena, Suse, Sven Manguard, Svenman, Svizra, Swapper, Swi521, Swid, Swiss Ace, SwissCelt, SwissTHX11384EB, Switzerlandinsound, Swordfish2008, Synchronism, Synthetik, Systemas, Szlam, Szwi, T.M.M. Dowd, TERIKATA, TFOWR, TGC55, THF, TLRkahuna, TUF-KAT, Ta bu shi da yu, Tad Lincoln, Taemyr, Tang Wenlong, Tangotango, Tanthalas39, Targeman, Tavis, Tbhotch, Tcc9mpl, Tedickey, TedlyW, Teemu Ruskeepää, Teerickson, Teiladnam, Tellsgeschoss, Tenmei, Tennis expert, Tequendamia, TerreOcre, TestPilot, Tetracube, Thadius856, Thatdog, The Egyptian Liberal, The Epopt, The Evil Spartan, The High Fin Sperm Whale, The Thing That Should Not Be, The Wild West guy, The Wordsmith, The sanch, The undertow, TheCatalyst31, TheKMan, TheProject, Theanthrope, Thecurran, Thedarxide, Thedjatclubrock, Thejason, Thelb4, Themanwithoutapast, Theonlytrandana, Therequiembellishere, Thingg, Think outside the box, ThomasK, Thomasankers, Thricecube, Thue, Tiddly Tom, Tide rolls, TiffaF, Tiggerjay, Tillwe, Tim bates, TimBentley, Timir2, Timlong28, Timrollpickering, Timwi, TippTopp, Tiptoety, Tirerim, Titoxd, Tjayh913, Tnxman307, Tobby72, Tobias Galtieri, Tobyc75, Tobyjhook, Todowd, Tofubar, Tohd8BohaithuGh1, Tom Morris, Tom Radulovich, TomH, TomPreuss, Tombomp, Tomchen1989, Tommy2010, Tony Fox, Tony1, Tonyle, Tosão, TouristPhilosopher, Toussaint, Tox, Tpbradbury, Trandana, TreasuryTag, Treemapper, Trigaranus, Trilobite, Triona, Trish Boland, TriviaKing, TrixKidRox, Trompeta, Trusilver, Tuomas, Turkuun, Twsx, Tye, UKe-CH, Uirauna, UkPaolo, Ukryan1992, Ulmke, Ulric1313, Ulritz, Ultimate Destiny, Uncle Milty, Uncle Scrooge, UncleDouggie, Unyoyega, User179514, Uusef, VI, VX-LTM, Vacekha, Van helsing, Vanished 6551232, Vanished user 579385839, Vardion, Vasilcho, Vcelloho, Veledan, Velella, Veritologist, Vettrock, Vianello, Victor falk, Victorjwilli, Vivisfugue, Vnsdvjnsvksnvjkmc, VolatileChemical, Vortexrealm, Voyager77377, Vrenator, Vsmith, Vzbs34, WTGDMan1986, WadeSimMiser, Wafulz, Waggers, Wakablogger2, Waldir, Walrus heart, Wangi, WarthogDemon, Washburnmav, Wayne Slam, Wbennettjr, Wbkelley, Welsh, Weregerbil, Wereon, WhisperToMe, Whoda, Whosyourjudas, Wik, WikHead, Wiki Raja, Wiki alf, Wiki-vr, Wiki.Tango.Foxtrot, WikiLaurent, Wikialthaf, Wikibob, Wikid77, Wikidemon, Wikidenizen, Wikieditor06, Wikiepdiax818, Wikievil666, Wikipedian Penguin, Wikipelli, Wikiscribe, Wikiturk, William Avery, William Rehtworc, Willking1979, Wimt, Wisdom89, Wmahan, WolfgangFaber, Wombat81, Woohookitty, World, Wormcast, Wprlh, Writtenright, X0man0x, X2wolf, Xaosflux, Xezbeth, Xonqnopp, Xyzzy n, Xyzzyva, Yaf, Yahia.barie, Yamamoto Ichiro, Yardbirdjo, Ybruf, YellowMonkey, Ynilp, Yorkshirian, Yvesnimmo, Ywmpq205, Z.E.R.O., Zabadinho, Zacharie Grossen, Zapmania, Zekerags, Zenosparadox, ZeppelinFan29, Zidonuke, Zntrip, Zoe, Zondor, ZooFari, Zscout370, Zurichtourismus, Zzuuzz, ÇŸ³å°è±◆, Île flottante, Александър, Гаджиев Азер, , 3799 anonymous edits

Image Sources, Licenses and Contributors

Image:Alter Strom.jpg *Source*: http://en.wikipedia.org/w/index.php?title=File:Alter_Strom.jpg *License*: unknown *Contributors*: User:Alabama, User:Fir0002

Image:IMG RoyalCanalnrKinnegad5706w.jpg *Source*: http://en.wikipedia.org/w/index.php?title=File:IMG_RoyalCanalnrKinnegad5706w.jpg *License*: unknown *Contributors*: Original uploader was Sarah777 at en.wikipedia

Image:MurtfaltalrDanubeChannel.JPG *Source*: http://en.wikipedia.org/w/index.php?title=File:MurtfaltalrDanubeChannel.JPG *License*: unknown *Contributors*: User:CristianChirita

Image:caen.hill.locks.in.devizes.arp.jpg *Source*: http://en.wikipedia.org/w/index.php?title=File:Caen.hill.locks.in.devizes.arp.jpg *License*: unknown *Contributors*: User:Arpingstone

Image:pontcysyllte aqueduct arp.jpg *Source*: http://en.wikipedia.org/w/index.php?title=File:Pontcysyllte_aqueduct_arp.jpg *License*: unknown *Contributors*: Adambro, Arpingstone, Grahamec, Tomer T

Image:Kaiserkanal01.jpg *Source*: http://en.wikipedia.org/w/index.php?title=File:Kaiserkanal01.jpg *License*: unknown *Contributors*: Airunp, Fb78, Ibn Battuta, Immanuel Giel, JackyR, Olivier2, Peter17, Ronaldino, Sanbec

File:Thal Canal.JPG *Source*: http://en.wikipedia.org/w/index.php?title=File:Thal_Canal.JPG *License*: unknown *Contributors*: Khalid Mahmood, Zaccarias, 2 anonymous edits

Image:Canal system in Lowell, Massachusetts.png *Source*: http://en.wikipedia.org/w/index.php?title=File:Canal_system_in_Lowell,_Massachusetts.png *License*: unknown *Contributors*: Mark M. Howland, Margy Chrisney

File:NYmohawk-ErieCanalRexford.JPG *Source*: http://en.wikipedia.org/w/index.php?title=File:NYmohawk-ErieCanalRexford.JPG *License*: unknown *Contributors*: Clifton Park Collection

Image:Canals USA 1825.png *Source*: http://en.wikipedia.org/w/index.php?title=File:Canals_USA_1825.png *License*: unknown *Contributors*: Allen Johnson

File:Viewing man-made canals in the marsh in Louisiana, Barataria Basin.jpg *Source*: http://en.wikipedia.org/w/index.php?title=File:Viewing_man-made_canals_in_the_marsh_in_Louisiana,_Barataria_Basin.jpg *License*: unknown *Contributors*: Dr. Terry McTigue, NOAA, NOS, ORR. Original uploader was Mattisse at en.wikipedia

Image:KeizersgrachtReguliersgrachtAmsterdam.jpg *Source*: http://en.wikipedia.org/w/index.php?title=File:KeizersgrachtReguliersgrachtAmsterdam.jpg *License*: unknown *Contributors*: User:Massimo Catarinella

File:Griboyedov Canal 2.jpg *Source*: http://en.wikipedia.org/w/index.php?title=File:Griboyedov_Canal_2.jpg *License*: unknown *Contributors*: Evgeny Gerashchenko

Image:Panama Canal Miraflores Locks.jpg *Source*: http://en.wikipedia.org/w/index.php?title=File:Panama_Canal_Miraflores_Locks.jpg *License*: unknown *Contributors*: User:Stan Shebs

Image:Amsterdamcannel333.JPG *Source*: http://en.wikipedia.org/w/index.php?title=File:Amsterdamcannel333.JPG *License*: unknown *Contributors*: Original uploader was Iijjccoo at en.wikipedia

File:Flag of Lombardy.svg *Source*: http://en.wikipedia.org/w/index.php?title=File:Flag_of_Lombardy.svg *License*: unknown *Contributors*: User:F l a n k e r

File:Lombardy in Italy.svg *Source*: http://en.wikipedia.org/w/index.php?title=File:Lombardy_in_Italy.svg *License*: unknown *Contributors*: DenghiùComm, TUBS

File:Lombardy.satellite.jpg *Source*: http://en.wikipedia.org/w/index.php?title=File:Lombardy.satellite.jpg *License*: unknown *Contributors*: Image Science and Analysis Laboratory, NASA-Johnson Space Center. 18 Mar. 2005. "Earth from Space - Image Information."

File:Orobie 5.jpg *Source*: http://en.wikipedia.org/w/index.php?title=File:Orobie_5.jpg *License*: unknown *Contributors*: AnRo0002, Paolo da Reggio

File:Le Lac Majeur.jpg *Source*: http://en.wikipedia.org/w/index.php?title=File:Le_Lac_Majeur.jpg *License*: unknown *Contributors*: Matalyn

File:ImperialpalaceMilan.jpg *Source*: http://en.wikipedia.org/w/index.php?title=File:ImperialpalaceMilan.jpg *License*: unknown *Contributors*: User:Lorenzo Fratti

File:CelPha.jpg *Source*: http://en.wikipedia.org/w/index.php?title=File:CelPha.jpg *License*: unknown *Contributors*: Cenomani

File:Nuremberg chronicles f 147v 1.jpg *Source*: http://en.wikipedia.org/w/index.php?title=File:Nuremberg_chronicles_f_147v_1.jpg *License*: unknown *Contributors*: Michel Wolgemut, Wilhelm Pleydenwurff (Text: Hartmann Schedel)

File:GGVCiSAdM.png *Source*: http://en.wikipedia.org/w/index.php?title=File:GGVCiSAdM.png *License*: unknown *Contributors*: Anovelo da Imbonate

File:Monza-palace.jpg *Source*: http://en.wikipedia.org/w/index.php?title=File:Monza-palace.jpg *License*: unknown *Contributors*:

File:Panorama of Milan from Palazzo Lombardia.jpg *Source*: http://en.wikipedia.org/w/index.php?title=File:Panorama_of_Milan_from_Palazzo_Lombardia.jpg *License*: unknown *Contributors*: User:Artemisi

File:San Martino della Battaglia Panorama 001.JPG *Source*: http://en.wikipedia.org/w/index.php?title=File:San_Martino_della_Battaglia_Panorama_001.JPG *License*: unknown *Contributors*: User:Janericloebe

File:PalazzoLombardia2.jpg *Source*: http://en.wikipedia.org/w/index.php?title=File:PalazzoLombardia2.jpg *License*: unknown *Contributors*: Conte di Cavour, Friedrichstrasse

File:Scena di duello R6 - Foppe - Nadro (Foto Luca Giarelli).jpg *Source*: http://en.wikipedia.org/w/index.php?title=File:Scena_di_duello_R6_-_Foppe_-_Nadro_(Foto_Luca_Giarelli).jpg *License*: unknown *Contributors*: Luca Giarelli

File:Villa olmo como pano.jpg *Source*: http://en.wikipedia.org/w/index.php?title=File:Villa_olmo_como_pano.jpg *License*: unknown *Contributors*: User:Alessio Facchin

File:Última Cena - Da Vinci 5.jpg *Source*: http://en.wikipedia.org/w/index.php?title=File:Última_Cena_-_Da_Vinci_5.jpg *License*: unknown *Contributors*: User:Platonides

File:Cotoletta e patate al forno.jpg *Source*: http://en.wikipedia.org/w/index.php?title=File:Cotoletta_e_patate_al_forno.jpg *License*: unknown *Contributors*: Paolo Piscolla

File:Teatro Grande (Sala).jpg *Source*: http://en.wikipedia.org/w/index.php?title=File:Teatro_Grande_(Sala).jpg *License*: unknown *Contributors*: User:Okkiproject

File:Lombardy Provinces.png *Source*: http://en.wikipedia.org/w/index.php?title=File:Lombardy_Provinces.png *License*: unknown *Contributors*: ALE!, Cflm001, Conscious, DenghiùComm, Gigillo83, Juiced lemon, Mortadelo2005, Vonvikken, ZooFari

File:Flag of Italy.svg *Source*: http://en.wikipedia.org/w/index.php?title=File:Flag_of_Italy.svg *License*: unknown *Contributors*: Anomie

File:Italy-Emblem.svg *Source*: http://en.wikipedia.org/w/index.php?title=File:Italy-Emblem.svg *License*: unknown *Contributors*: User:F l a n k e r

File:EU-Italy.svg *Source*: http://en.wikipedia.org/w/index.php?title=File:EU-Italy.svg *License*: unknown *Contributors*: User:NuclearVacuum

File:Increase2.svg *Source*: http://en.wikipedia.org/w/index.php?title=File:Increase2.svg *License*: unknown *Contributors*: Sarang

Image:Speakerlink.svg *Source*: http://en.wikipedia.org/w/index.php?title=File:Speakerlink.svg *License*: unknown *Contributors*: Woodstone. Original uploader was Woodstone at en.wikipedia

File:Colosseum in Rome, Italy - April 2007.jpg *Source*: http://en.wikipedia.org/w/index.php?title=File:Colosseum_in_Rome,_Italy_-_April_2007.jpg *License*: unknown *Contributors*: User:Diliff

File:Naval Jack of Italy.svg *Source*: http://en.wikipedia.org/w/index.php?title=File:Naval_Jack_of_Italy.svg *License*: unknown *Contributors*: User:Denelson83

File:Vitruvian.jpg *Source*: http://en.wikipedia.org/w/index.php?title=File:Vitruvian.jpg *License*: unknown *Contributors*: Czarnoglowa, Davepape, Dbenbenn, DenghiùComm, Editor at Large, Eleassar, Flominator, G.dallorto, Gepardenforellenfischer, Holek, Jeff Dahl, Jeff G., Jon Harald Søby, Jökullinn, LadyInGrey, Lennert B, LoStrangolatore, Locutus Borg, Mdd, Mglanznig, OldakQuill, Patrick, Ranveig, Rocket000, Sbwoodside, Simonizer, Sven, Szczepan1990, Turzh, , 12 anonymous edits

File:Lejeune - Bataille de Marengo.jpg *Source*: http://en.wikipedia.org/w/index.php?title=File:Lejeune_-_Bataille_de_Marengo.jpg *License*: unknown *Contributors*: Arachn0, Brianboulton, Bukk, Coyau, Frank Schulenburg, Kilom691, Man vyi, Mutter Erde

File:With Victor Emmanuel.jpg *Source*: http://en.wikipedia.org/w/index.php?title=File:With_Victor_Emmanuel.jpg *License*: unknown *Contributors*: Sebastiano De Albertis (1828-1897)

File:Italia-I-Guerra-Mondiale-fanti-classe-1888.jpg *Source*: http://en.wikipedia.org/w/index.php?title=File:Italia-I-Guerra-Mondiale-fanti-classe-1888.jpg *License*: unknown *Contributors*: User:Albertomos

File:Hitler and Mussolini June 1940.jpg *Source*: http://en.wikipedia.org/w/index.php?title=File:Hitler_and_Mussolini_June_1940.jpg *License*: unknown *Contributors*: Eva Braun

File:Costituzione della Repubblica Italiana.jpg *Source*: http://en.wikipedia.org/w/index.php?title=File:Costituzione_della_Repubblica_Italiana.jpg *License*: unknown *Contributors*: Figiu, Gusme

File:Treaty of Rome.jpg *Source*: http://en.wikipedia.org/w/index.php?title=File:Treaty_of_Rome.jpg *License*: unknown *Contributors*: ANGELUS, Pieter Kuiper, 1 anonymous edits

File:Italy topographic map-blank.svg *Source*: http://en.wikipedia.org/w/index.php?title=File:Italy_topographic_map-blank.svg *License*: unknown *Contributors*: User:Sting

File:Mont Blanc from Aosta Valley.JPG *Source*: http://en.wikipedia.org/w/index.php?title=File:Mont_Blanc_from_Aosta_Valley.JPG *License*: unknown *Contributors*: User:Ximonic

File:National parks of Italy.svg *Source*: http://en.wikipedia.org/w/index.php?title=File:National_parks_of_Italy.svg *License*: unknown *Contributors*: ColdShine, 2 anonymous edits

File:Torre della pelosa + isola piana + asinara da capo falcone.jpg *Source*: http://en.wikipedia.org/w/index.php?title=File:Torre_della_pelosa_+_isola_piana_+_asinara_da_capo_falcone.jpg *License*: unknown *Contributors*: enrico_sirola

File:Vernazza.JPG *Source*: http://en.wikipedia.org/w/index.php?title=File:Vernazza.JPG *License*: unknown *Contributors*: AnticheSere

File:Italian regions provinces white no labels.svg *Source*: http://en.wikipedia.org/w/index.php?title=File:Italian_regions_provinces_white_no_labels.svg *License*: unknown *Contributors*: User:Kat888

File:Giorgio Napolitano-M.jpg *Source*: http://en.wikipedia.org/w/index.php?title=File:Giorgio_Napolitano-M.jpg *License*: unknown *Contributors*: Cruccone, EPO, G.dallorto, Krinkle, Myworld

File:Aula Montecitorio.jpg *Source*: http://en.wikipedia.org/w/index.php?title=File:Aula_Montecitorio.jpg *License*: unknown *Contributors*: Original uploader was RuandaBit at it.wikipedia

File:Roma-palazzaccio.jpg *Source*: http://en.wikipedia.org/w/index.php?title=File:Roma-palazzaccio.jpg *License*: unknown *Contributors*: FlickreviewR, G.dallorto, Lalupa, Mac9

File:Obama Napolitano.jpg *Source*: http://en.wikipedia.org/w/index.php?title=File:Obama_Napolitano.jpg *License*: unknown *Contributors*: Presidenza della Repubblica Italiana

File:Cavour (550).jpg *Source*: http://en.wikipedia.org/w/index.php?title=File:Cavour_(550).jpg *License*: unknown *Contributors*: User:Gaetano56

File:Eurofighter Typhoon 02.jpg *Source*: http://en.wikipedia.org/w/index.php?title=File:Eurofighter_Typhoon_02.jpg *License*: unknown *Contributors*: User:Dmottl

File:2012-03-07 Motorshow Geneva 4309.JPG *Source*: http://en.wikipedia.org/w/index.php?title=File:2012-03-07_Motorshow_Geneva_4309.JPG *License*: unknown *Contributors*: User:Noebu

File:Eurozone.svg *Source*: http://en.wikipedia.org/w/index.php?title=File:Eurozone.svg *License*: unknown *Contributors*: User:Glentamara

File:Vineyards in Piemonte, Italy.jpg *Source*: http://en.wikipedia.org/w/index.php?title=File:Vineyards_in_Piemonte,_Italy.jpg *License*: unknown *Contributors*: Megan Mallen

File:Frecciarossa.JPG *Source*: http://en.wikipedia.org/w/index.php?title=File:Frecciarossa.JPG *License*: unknown *Contributors*: Original uploader was StuporesMundi at it.wikipedia

File:Popolazione Italia 1861 2008.jpg *Source*: http://en.wikipedia.org/w/index.php?title=File:Popolazione_Italia_1861_2008.jpg *License*: unknown *Contributors*: T,C&S. Original uploader was T,C&S at it.wikipedia

File:Mulberry Street NYC c1900 LOC 3g04637u edit.jpg *Source*: http://en.wikipedia.org/w/index.php?title=File:Mulberry_Street_NYC_c1900_LOC_3g04637u_edit.jpg *License*: unknown *Contributors*: Original uploader was Debivort at en.wikipedia

File:COB data Italy.PNG *Source*: http://en.wikipedia.org/w/index.php?title=File:COB_data_Italy.PNG *License*: unknown *Contributors*: User:Kransky

File:Map Italophone World - updated.png *Source*: http://en.wikipedia.org/w/index.php?title=File:Map_Italophone_World_-_updated.png *License*: unknown *Contributors*: Fsolda, Otourly, Sinigagl, 4 anonymous edits

File:Santa Maria del Fiore.jpg *Source*: http://en.wikipedia.org/w/index.php?title=File:Santa_Maria_del_Fiore.jpg *License*: unknown *Contributors*: Andreagrossmann, Bouncey2k, G.dallorto, Gryffindor, Sailko

File:Université La Sapienza.JPG *Source*: http://en.wikipedia.org/w/index.php?title=File:Université_La_Sapienza.JPG *License*: unknown *Contributors*: User:LPLT

File:The Archiginnasio, Bologna, Italy, the wing with the Anatomical theatre.JPG *Source*: http://en.wikipedia.org/w/index.php?title=File:The_Archiginnasio,_Bologna,_Italy,_the_wing_with_the_Anatomical_theatre.JPG *License*: unknown *Contributors*: User:Luca Borghi

File:Tessera sanitaria.png *Source*: http://en.wikipedia.org/w/index.php?title=File:Tessera_sanitaria.png *License*: unknown *Contributors*: User:Balfabio

File:Venezia-punta della dogana.jpg *Source*: http://en.wikipedia.org/w/index.php?title=File:Venezia-punta_della_dogana.jpg *License*: unknown *Contributors*: FlickreviewR, G.dallorto, Mac9

File:Pisa Duomo.jpg *Source*: http://en.wikipedia.org/w/index.php?title=File:Pisa_Duomo.jpg *License*: unknown *Contributors*: User:Lucarelli

File:Leonardo da Vinci (1452-1519) - The Last Supper (1495-1498).jpg *Source*: http://en.wikipedia.org/w/index.php?title=File:Leonardo_da_Vinci_(1452-1519)_-_The_Last_Supper_(1495-1498).jpg *License*: unknown *Contributors*: Alvaro qc, AndreasPraefcke, Avatar, Blurpeace, Czarnoglowa, Durova, Emijrp, G.dallorto, Jarekt, OldakQuill, PeterSymonds, Schaengel89, TTaylor, The Evil IP address, Thebrid, Vonvon, 27 anonymous edits

File:DanteDetail.jpg *Source*: http://en.wikipedia.org/w/index.php?title=File:DanteDetail.jpg *License*: unknown *Contributors*: Wetman

File:GiacomoPuccini.jpg *Source*: http://en.wikipedia.org/w/index.php?title=File:GiacomoPuccini.jpg *License*: unknown *Contributors*: Original uploader was NewYork1956 at en.wikipedia

File:Luciano Pavarotti 15.06.02 cropped2.jpg *Source*: http://en.wikipedia.org/w/index.php?title=File:Luciano_Pavarotti_15.06.02_cropped2.jpg *License*: unknown *Contributors*: Pirlouiiiit from Marseille, France

File:Federico Fellini NYWTS 2.jpg *Source*: http://en.wikipedia.org/w/index.php?title=File:Federico_Fellini_NYWTS_2.jpg *License*: unknown *Contributors*: Walter Albertin, World Telegram staff photographer

File:Justus Sustermans - Portrait of Galileo Galilei, 1636.jpg *Source*: http://en.wikipedia.org/w/index.php?title=File:Justus_Sustermans_-_Portrait_of_Galileo_Galilei,_1636.jpg *License*: unknown *Contributors*: Alno, Dmitry Rozhkov, G.dallorto, Lupo, Meno25, Myself488, Phrood, Ragesoss, Sercan.ergün, Shakko, Wutsje, 15 anonymous edits

File:Scudo2009.jpg *Source*: http://en.wikipedia.org/w/index.php?title=File:Scudo2009.jpg *License*: unknown *Contributors*: oscar federico bodini

File:Milan Fashion Week 1.jpg *Source*: http://en.wikipedia.org/w/index.php?title=File:Milan_Fashion_Week_1.jpg *License*: unknown *Contributors*: br1dotcom

File:Eq it-na pizza-margherita sep2005 sml-2.png *Source*: http://en.wikipedia.org/w/index.php?title=File:Eq_it-na_pizza-margherita_sep2005_sml-2.png *License*: unknown *Contributors*: User:Itzuvit

File:Openstreetmap logo.svg *Source*: http://en.wikipedia.org/w/index.php?title=File:Openstreetmap_logo.svg *License*: unknown *Contributors*: OpenStreetMap

File:Pavia Ponte Coperto.JPG *Source*: http://en.wikipedia.org/w/index.php?title=File:Pavia_Ponte_Coperto.JPG *License*: unknown *Contributors*: Welleschik

File:Ticino at Gotthard-Tunnel.jpg *Source*: http://en.wikipedia.org/w/index.php?title=File:Ticino_at_Gotthard-Tunnel.jpg *License*: unknown *Contributors*: User:Hans G. Oberlack

File:Collage Pavia.jpg *Source*: http://en.wikipedia.org/w/index.php?title=File:Collage_Pavia.jpg *License*: unknown *Contributors*: User:Lolo00

File:Pavia-Stemma.png *Source*: http://en.wikipedia.org/w/index.php?title=File:Pavia-Stemma.png *License*: unknown *Contributors*: DenghiùComm, G.dallorto, Idéfix, Magul, MauroVan, Oxhop

file:Italy provincial location map.svg *Source*: http://en.wikipedia.org/w/index.php?title=File:Italy_provincial_location_map.svg *License*: unknown *Contributors*: TUBS

File:Red pog.svg *Source*: http://en.wikipedia.org/w/index.php?title=File:Red_pog.svg *License*: unknown *Contributors*: Anomie

File:Pavia Certosa 1.jpg *Source*: http://en.wikipedia.org/w/index.php?title=File:Pavia_Certosa_1.jpg *License*: unknown *Contributors*: G.dallorto, Mac9, Marco Bonavoglia

File:San michele maggiore.JPG *Source*: http://en.wikipedia.org/w/index.php?title=File:San_michele_maggiore.JPG *License*: unknown *Contributors*: Matteo Agostinelli

File:Castello Visconteo (Pavia).JPG *Source*: http://en.wikipedia.org/w/index.php?title=File:Castello_Visconteo_(Pavia).JPG *License*: unknown *Contributors*: Giorgio Gonnella

File:Santa maria del carmine.jpg *Source*: http://en.wikipedia.org/w/index.php?title=File:Santa_maria_del_carmine.jpg *License*: unknown *Contributors*: Matteo Agostinelli

File:Pavia 3.jpg *Source*: http://en.wikipedia.org/w/index.php?title=File:Pavia_3.jpg *License*: unknown *Contributors*: Attilios, G.dallorto, Ian Spackman

File:Speaker Icon.svg *Source*: http://en.wikipedia.org/w/index.php?title=File:Speaker_Icon.svg *License*: unknown *Contributors*: Blast, G.Hagedorn, Mobius, 2 anonymous edits

File:Collegio Borromeo (Pavia).jpg *Source*: http://en.wikipedia.org/w/index.php?title=File:Collegio_Borromeo_(Pavia).jpg *License*: unknown *Contributors*: Giorgio Gonnella

File:Aula magna-University-Pavia-Italy.jpg *Source*: http://en.wikipedia.org/w/index.php?title=File:Aula_magna-University-Pavia-Italy.jpg *License*: unknown *Contributors*: Giorgio Gonnella

File:Flag of Europe.svg *Source*: http://en.wikipedia.org/w/index.php?title=File:Flag_of_Europe.svg *License*: unknown *Contributors*: User:-xfi-, User:Dbenbenn, User:Funakoshi, User:Jeltz, User:Nightstallion, User:Paddu, User:Verdy p, User:Zscout370

File:Flag of San Marino.svg *Source*: http://en.wikipedia.org/w/index.php?title=File:Flag_of_San_Marino.svg *License*: unknown *Contributors*: User:Zscout370

File:Flag of the Sovereign Military Order of Malta.svg *Source*: http://en.wikipedia.org/w/index.php?title=File:Flag_of_the_Sovereign_Military_Order_of_Malta.svg *License*: unknown *Contributors*: Zscout370

File:Flag of Switzerland.svg *Source*: http://en.wikipedia.org/w/index.php?title=File:Flag_of_Switzerland.svg *License*: unknown *Contributors*: User:-xfi-, User:Marc Mongenet, User:Zscout370

File:Flag of the Vatican City.svg *Source*: http://en.wikipedia.org/w/index.php?title=File:Flag_of_the_Vatican_City.svg *License*: unknown *Contributors*: AguaitantPV, Consta, DenghiùComm, Denniss, Durin, F l a n k e r, Fry1989, Homo lupus, Jarekt, Jed, Klemen Kocjancic, Krun, Liftarn, Ludger1961, Mattes, Next2u, NielsF, Nightstallion, OAlexander, Pumbaa80, Str4nd, Zscout370, 13 anonymous edits

File:Flag of Slovenia.svg *Source*: http://en.wikipedia.org/w/index.php?title=File:Flag_of_Slovenia.svg *License*: unknown *Contributors*: User:SKopp, User:Vzb83, User:Zscout370

File:Flag of Croatia.svg *Source*: http://en.wikipedia.org/w/index.php?title=File:Flag_of_Croatia.svg *License*: unknown *Contributors*: User:MaGa

File:Flag of Somalia.svg *Source*: http://en.wikipedia.org/w/index.php?title=File:Flag_of_Somalia.svg *License*: unknown *Contributors*: Anakin101, Anime Addict AA, Anonymous101, Flad, Fry1989, Gdgourou, Homo lupus, JLogan, Klemen Kocjancic, Mattes, Permjak, SKopp, ThomasPusch, Trijnstel, Tvdm, Ultratomio, Vzb83, احمد.عطية1, 12 anonymous edits

File:Flag of Eritrea.svg *Source*: http://en.wikipedia.org/w/index.php?title=File:Flag_of_Eritrea.svg *License*: unknown *Contributors*: Alkari, Bukk, Counny, Fry1989, Homo lupus, Klemen Kocjancic, Mattes, Moipaulochon, Neq00, Nightstallion, Ninane, Persiana, Ratatosk, Rodejong, ThomasPusch, Vzb83, WikipediaMaster, Zscout370, 2 anonymous edits

File:Flag of Libya.svg *Source*: http://en.wikipedia.org/w/index.php?title=File:Flag_of_Libya.svg *License*: unknown *Contributors*: Various

File:Flag of Malta.svg *Source*: http://en.wikipedia.org/w/index.php?title=File:Flag_of_Malta.svg *License*: unknown *Contributors*: Alkari, Fry1989, Gabbe, Homo lupus, Klemen Kocjancic, Liftarn, Mattes, Meno25, Nightstallion, Peeperman, Pumbaa80, Ratatosk, Rodejong, Zscout370, 4 anonymous edits

File:Flag of Kosovo.svg *Source*: http://en.wikipedia.org/w/index.php?title=File:Flag_of_Kosovo.svg *License*: unknown *Contributors*: User:Cradel, User:Ningyou

File:Flag of Montenegro.svg *Source*: http://en.wikipedia.org/w/index.php?title=File:Flag_of_Montenegro.svg *License*: unknown *Contributors*: User:B1mbo, User:Froztbyte

File:Knowledge Italian Europe map.jpg *Source*: http://en.wikipedia.org/w/index.php?title=File:Knowledge_Italian_Europe_map.jpg *License*: unknown *Contributors*: Hidra92 Modified from original version by and uploaded on it.wiki with same name.

File:Map Italophone World.png *Source*: http://en.wikipedia.org/w/index.php?title=File:Map_Italophone_World.png *License*: unknown *Contributors*: User:Ilario

File:Lengua italiana.png *Source*: http://en.wikipedia.org/w/index.php?title=File:Lengua_italiana.png *License*: unknown *Contributors*: User:Fobos92

File:Flag of France.svg *Source*: http://en.wikipedia.org/w/index.php?title=File:Flag_of_France.svg *License*: unknown *Contributors*: Anomie

File:Flag of Albania.svg *Source*: http://en.wikipedia.org/w/index.php?title=File:Flag_of_Albania.svg *License*: unknown *Contributors*: User:Dbenbenn

File:Flag of Monaco.svg *Source*: http://en.wikipedia.org/w/index.php?title=File:Flag_of_Monaco.svg *License*: unknown *Contributors*: User:SKopp

File:Flag of Greece.svg *Source*: http://en.wikipedia.org/w/index.php?title=File:Flag_of_Greece.svg *License*: unknown *Contributors*: (of code) (talk)

File:Flag of Ukraine.svg *Source*: http://en.wikipedia.org/w/index.php?title=File:Flag_of_Ukraine.svg *License*: unknown *Contributors*: User:Jon Harald Søby, User:Zscout370

File:Flag of Italy (1861-1946) crowned.svg *Source*: http://en.wikipedia.org/w/index.php?title=File:Flag_of_Italy_(1861-1946)_crowned.svg *License*: unknown *Contributors*: User:F l a n k e r

File:Free Territory Trieste Flag.svg *Source*: http://en.wikipedia.org/w/index.php?title=File:Free_Territory_Trieste_Flag.svg *License*: unknown *Contributors*: Original uploader was Kuemmjen at de.wikipedia

File:Flag of the Free State of Fiume.svg *Source*: http://en.wikipedia.org/w/index.php?title=File:Flag_of_the_Free_State_of_Fiume.svg *License*: unknown *Contributors*: Mysid

File:Flag of Ethiopia.svg *Source*: http://en.wikipedia.org/w/index.php?title=File:Flag_of_Ethiopia.svg *License*: unknown *Contributors*: Aaker, Anime Addict AA, Antemister, Cycn, F l a n k e r, Fry1989, GoodMorningEthiopia, Happenstance, Homo lupus, Huhsunqu, Ixfd64, Klemen Kocjancic, MartinThoma, Mattes, Mozzan, Neq00, OAlexander, Pumbaa80, Rainforest tropicana, Reisio, SKopp, Smooth O, Spiritia, ThomasPusch, Torstein, Wsiegmund, Zscout370, 16 anonymous edits

File:Flag of Cyprus.svg *Source*: http://en.wikipedia.org/w/index.php?title=File:Flag_of_Cyprus.svg *License*: unknown *Contributors*: User:Vzb83

File:Flag of the People's Republic of China.svg *Source*: http://en.wikipedia.org/w/index.php?title=File:Flag_of_the_People's_Republic_of_China.svg *License*: unknown *Contributors*: User:Denelson83, User:SKopp, User:Shizhao, User:Zscout370

File:Flag of Egypt.svg *Source*: http://en.wikipedia.org/w/index.php?title=File:Flag_of_Egypt.svg *License*: unknown *Contributors*: Open Clip Art

File:Flag of Tunisia.svg *Source*: http://en.wikipedia.org/w/index.php?title=File:Flag_of_Tunisia.svg *License*: unknown *Contributors*: Alkari, AnonMoos, Avala, Bender235, Duduziq, Elina2308, Emmanuel.boutet, Flad, Fry1989, Gabbe, Juiced lemon, Klemen Kocjancic, Mattes, Meno25, Moumou82, Myself488, Neq00, Nightstallion, Reisio, Str4nd, TFCforever, Ö, Фёдор Гусляров, 9 anonymous edits

File:Flag of Austria-Hungary 1869-1918.svg *Source*: http://en.wikipedia.org/w/index.php?title=File:Flag_of_Austria-Hungary_1869-1918.svg *License*: unknown *Contributors*: User:B1mbo

File:Flag of Turkey.svg *Source*: http://en.wikipedia.org/w/index.php?title=File:Flag_of_Turkey.svg *License*: unknown *Contributors*: User:Dbenbenn

File:Flag of Argentina.svg *Source*: http://en.wikipedia.org/w/index.php?title=File:Flag_of_Argentina.svg *License*: unknown *Contributors*: User:Dbenbenn

File:Flag of the United States.svg *Source*: http://en.wikipedia.org/w/index.php?title=File:Flag_of_the_United_States.svg *License*: unknown *Contributors*: Anomie

File:Flag of Canada.svg *Source*: http://en.wikipedia.org/w/index.php?title=File:Flag_of_Canada.svg *License*: unknown *Contributors*: Anomie

File:Flag of Germany.svg *Source*: http://en.wikipedia.org/w/index.php?title=File:Flag_of_Germany.svg *License*: unknown *Contributors*: Anomie

File:Flag of Venezuela.svg *Source*: http://en.wikipedia.org/w/index.php?title=File:Flag_of_Venezuela.svg *License*: unknown *Contributors*: Alkari, Bastique, Denelson83, DerFussi, Fry1989, George McFinnigan, Herbythyme, Homo lupus, Huhsunqu, Infrogmation, K21edgo, Klemen Kocjancic, Ludger1961, Neq00, Nightstallion, Reisio, Rupert Pupkin, ThomasPusch, Vzb83, Wikisole, Zscout370, 12 anonymous edits

File:Flag of Australia.svg *Source*: http://en.wikipedia.org/w/index.php?title=File:Flag_of_Australia.svg *License*: unknown *Contributors*: Anomie, Mifter

File:Flag of Belgium (civil).svg *Source*: http://en.wikipedia.org/w/index.php?title=File:Flag_of_Belgium_(civil).svg *License*: unknown *Contributors*: Bean49, David Descamps, Dbenbenn, Denelson83, Evanc0912, Fry1989, Gabriel trzy, Howcome, IvanOS, Ms2ger, Nightstallion, Oreo Priest, Rocket000, Rodejong, Sir Iain, ThomasPusch, Warddr, Zscout370, 4 anonymous edits

File:Flag of Mexico.svg *Source*: http://en.wikipedia.org/w/index.php?title=File:Flag_of_Mexico.svg *License*: unknown *Contributors*: User:AlexCovarrubias

File:Flag of the United Kingdom.svg *Source*: http://en.wikipedia.org/w/index.php?title=File:Flag_of_the_United_Kingdom.svg *License*: unknown *Contributors*: Anomie, Mifter

File:Flag of Uruguay.svg *Source*: http://en.wikipedia.org/w/index.php?title=File:Flag_of_Uruguay.svg *License*: unknown *Contributors*: Alkari, CommonsDelinker, Fry1989, Homo lupus, Huhsunqu, Kineto007, Klemen Kocjancic, Kookaburra, Lorakesz, Mattes, Neq00, Nightstallion, Pumbaa80, Reisio, ThomasPusch, Zscout370, , 8 anonymous edits

File:Flag of Brazil.svg *Source*: http://en.wikipedia.org/w/index.php?title=File:Flag_of_Brazil.svg *License*: unknown *Contributors*: Anomie

File:Flag of Colombia.svg *Source*: http://en.wikipedia.org/w/index.php?title=File:Flag_of_Colombia.svg *License*: unknown *Contributors*: User:SKopp

File:Italian languages.png *Source*: http://en.wikipedia.org/w/index.php?title=File:Italian_languages.png *License*: unknown *Contributors*: User:Susana Freixeiro, User:XaneZeggi

File:Collage Venezia.jpg *Source*: http://en.wikipedia.org/w/index.php?title=File:Collage_Venezia.jpg *License*: unknown *Contributors*: User:DaniDF1995

File:StemmaVene.PNG *Source*: http://en.wikipedia.org/w/index.php?title=File:StemmaVene.PNG *License*: unknown *Contributors*: User:Daufer

File:Rialto Bridge Grand Canal.jpg *Source*: http://en.wikipedia.org/w/index.php?title=File:Rialto_Bridge_Grand_Canal.jpg *License*: unknown *Contributors*: User:Saffron Blaze

File:P1060341.jpg *Source*: http://en.wikipedia.org/w/index.php?title=File:P1060341.jpg *License*: unknown *Contributors*: Skyguy414

File:San Marco horses.jpg *Source*: http://en.wikipedia.org/w/index.php?title=File:San_Marco_horses.jpg *License*: unknown *Contributors*: Errabee, G.dallorto, Gveret Tered, Mac9, Man vyi, Olivier2, Thuresson

File:venice09.jpg *Source*: http://en.wikipedia.org/w/index.php?title=File:Venice09.jpg *License*: unknown *Contributors*: G.dallorto, Kkkdc, Shyam

File:The Grand Canal, Venice c1760 Francesco Guardi.jpg *Source*: http://en.wikipedia.org/w/index.php?title=File:The_Grand_Canal,_Venice_c1760_Francesco_Guardi.jpg *License*: unknown *Contributors*: AndreasPraefcke, Bukk, Krscal, Zolo

File:Venice by Piri Reis.jpg *Source*: http://en.wikipedia.org/w/index.php?title=File:Venice_by_Piri_Reis.jpg *License*: unknown *Contributors*: Original uploader was Shuppiluliuma at en.wikipedia

File:Venice, by Bolognino Zaltieri, 1565.jpg *Source*: http://en.wikipedia.org/w/index.php?title=File:Venice,_by_Bolognino_Zaltieri,_1565.jpg *License*: unknown *Contributors*: G. Braun (1541-1622) and F. Hogenberg(1535-1590), from the engravings by Bolognino Zaltieri (fl. 1560-80)

File:Venezia-map 1-1220x900.png *Source*: http://en.wikipedia.org/w/index.php?title=File:Venezia-map_1-1220x900.png *License*: unknown *Contributors*: User:Luestling

File:Venezia acqua alta notte 2005 modificata.jpg *Source*: http://en.wikipedia.org/w/index.php?title=File:Venezia_acqua_alta_notte_2005_modificata.jpg *License*: unknown *Contributors*: Paolo da Reggio

File:Venice.longshot.981pix (1).jpg *Source*: http://en.wikipedia.org/w/index.php?title=File:Venice.longshot.981pix_(1).jpg *License*: unknown *Contributors*: Image courtesy NASA/GSFC/MITI/ERSDAC/JAROS, and U.S./Japan ASTER Science Team Original uploader was Arpingstone at en.wikipedia

File:Veneto Venezia1 tango7174.jpg *Source*: http://en.wikipedia.org/w/index.php?title=File:Veneto_Venezia1_tango7174.jpg *License*: unknown *Contributors*: User:Tango7174

File:Sestieri di Venezia.svg *Source*: http://en.wikipedia.org/w/index.php?title=File:Sestieri_di_Venezia.svg *License*: unknown *Contributors*: Giovanni Fasano

File:Venezia Aerial View.jpg *Source*: http://en.wikipedia.org/w/index.php?title=File:Venezia_Aerial_View.jpg *License*: unknown *Contributors*: User:Unofeld781

File:Venice IMG 3731.JPG *Source*: http://en.wikipedia.org/w/index.php?title=File:Venice_IMG_3731.JPG *License*: unknown *Contributors*: Владимир Шеляпин

File:Venedig Seufzerbrücke.jpg *Source*: http://en.wikipedia.org/w/index.php?title=File:Venedig_Seufzerbrücke.jpg *License*: unknown *Contributors*: User:BertS

File:Venice Gondola Grand Canal.jpg *Source*: http://en.wikipedia.org/w/index.php?title=File:Venice_Gondola_Grand_Canal.jpg *License*: unknown *Contributors*: User:Saffron Blaze

File:Venice as seen from the air with bridge to mainland.jpg *Source*: http://en.wikipedia.org/w/index.php?title=File:Venice_as_seen_from_the_air_with_bridge_to_mainland.jpg *License*: unknown *Contributors*: Chris 73, Sitacuisses

File:Morning Impression along a Canal in Venice.JPG *Source*: http://en.wikipedia.org/w/index.php?title=File:Morning_Impression_along_a_Canal_in_Venice.JPG *License*: unknown *Contributors*: Rafail Sergeevich Levitsky

File:MunicipalitaVCE.png *Source*: http://en.wikipedia.org/w/index.php?title=File:MunicipalitaVCE.png *License*: unknown *Contributors*: it:user:Frassionsistematiche

File:Venezia-DSCF9752.JPG *Source*: http://en.wikipedia.org/w/index.php?title=File:Venezia-DSCF9752.JPG *License*: unknown *Contributors*: User:Rinina25, User:Twice25

File:PI5D9E~2Ca' Rezzonico.JPG *Source*: http://en.wikipedia.org/w/index.php?title=File:PI5D9E~2Ca'_Rezzonico.JPG *License*: unknown *Contributors*: User:G.Lanting

File:Teatro-la-fenice-sala.jpg *Source*: http://en.wikipedia.org/w/index.php?title=File:Teatro-la-fenice-sala.jpg *License*: unknown *Contributors*: Albertomos, Alensha, Bohème, DEDB, Gryffindor, Mac9, Pavel Krok, Via null

File:Rialto IMG 3996.JPG *Source*: http://en.wikipedia.org/w/index.php?title=File:Rialto_IMG_3996.JPG *License*: unknown *Contributors*: User:Deror avi

File:Pietro Longhi 025.jpg *Source*: http://en.wikipedia.org/w/index.php?title=File:Pietro_Longhi_025.jpg *License*: unknown *Contributors*: AndreasPraefcke, Anna reg, Bukk, G.dallorto, Wst

File:Casanova ritratto.jpg *Source*: http://en.wikipedia.org/w/index.php?title=File:Casanova_ritratto.jpg *License*: unknown *Contributors*: Airunp, Annenkov, Ecummenic, Jdsteakley, Red devil 666, Wst, 1 anonymous edits

File:Aldus-Horace2.jpeg *Source*: http://en.wikipedia.org/w/index.php?title=File:Aldus-Horace2.jpeg *License*: unknown *Contributors*: Aristeas, Daniel Baránek, Garabombo

File:Canaletto (II) 002.jpg *Source*: http://en.wikipedia.org/w/index.php?title=File:Canaletto_(II)_002.jpg *License*: unknown *Contributors*: Abxbay, BlackIceNRW, EDUCA33E, G.dallorto, Jacklee, Warburg, Zolo, 1 anonymous edits

File:Glassvenetian.jpg *Source*: http://en.wikipedia.org/w/index.php?title=File:Glassvenetian.jpg *License*: unknown *Contributors*: VAwebteam

File:Palazzo compagni, salone, lampadario di murano 01.JPG *Source*: http://en.wikipedia.org/w/index.php?title=File:Palazzo_compagni,_salone,_lampadario_di_murano_01.JPG *License*: unknown *Contributors*: user:sailko

File:Flag of Iran.svg *Source*: http://en.wikipedia.org/w/index.php?title=File:Flag_of_Iran.svg *License*: unknown *Contributors*: Various

File:Flag of Pakistan.svg *Source*: http://en.wikipedia.org/w/index.php?title=File:Flag_of_Pakistan.svg *License*: unknown *Contributors*: Zscout370

File:Flag of Indonesia.svg *Source*: http://en.wikipedia.org/w/index.php?title=File:Flag_of_Indonesia.svg *License*: unknown *Contributors*: User:Gabbe, User:SKopp

File:Flag of Thailand.svg *Source*: http://en.wikipedia.org/w/index.php?title=File:Flag_of_Thailand.svg *License*: unknown *Contributors*: User:Zscout370

File:Flag of Estonia.svg *Source*: http://en.wikipedia.org/w/index.php?title=File:Flag_of_Estonia.svg *License*: unknown *Contributors*: User:PeepP, User:SKopp

File:Flag of Bosnia and Herzegovina.svg *Source*: http://en.wikipedia.org/w/index.php?title=File:Flag_of_Bosnia_and_Herzegovina.svg *License*: unknown *Contributors*: User:Kseferovic

File:Flag of Armenia.svg *Source*: http://en.wikipedia.org/w/index.php?title=File:Flag_of_Armenia.svg *License*: unknown *Contributors*: User:SKopp

File:Flag of Portugal.svg *Source*: http://en.wikipedia.org/w/index.php?title=File:Flag_of_Portugal.svg *License*: unknown *Contributors*: User:Nightstallion

File:Flag of Russia.svg *Source*: http://en.wikipedia.org/w/index.php?title=File:Flag_of_Russia.svg *License*: unknown *Contributors*: Anomie

File:Flag of the Czech Republic.svg *Source*: http://en.wikipedia.org/w/index.php?title=File:Flag_of_the_Czech_Republic.svg *License*: unknown *Contributors*: special commission (of code): SVG version by cs:-xfi-. Colors according to Appendix No. 3 of czech legal Act 3/1993. cs:Zirland.

File:Flag of Switzerland (Pantone).svg *Source*: http://en.wikipedia.org/w/index.php?title=File:Flag_of_Switzerland_(Pantone).svg *License*: unknown *Contributors*: User:-xfi-, User:Marc Mongenet, User:Zscout370

File:Coat of Arms of Switzerland (Pantone).svg *Source*: http://en.wikipedia.org/w/index.php?title=File:Coat_of_Arms_of_Switzerland_(Pantone).svg *License*: unknown *Contributors*: User:E Pluribus Anthony, User:Reisio, User:Wiki-vr, User:Zscout370

File:Europe-Switzerland.svg *Source*: http://en.wikipedia.org/w/index.php?title=File:Europe-Switzerland.svg *License*: unknown *Contributors*: User:Hayden120, User:NuclearVacuum

File:Theater Kaiseraugst.jpg *Source*: http://en.wikipedia.org/w/index.php?title=File:Theater_Kaiseraugst.jpg *License*: unknown *Contributors*: User:Taxiarchos228

File:Bundesbrief.jpg *Source*: http://en.wikipedia.org/w/index.php?title=File:Bundesbrief.jpg *License*: unknown *Contributors*: AnRo0002, AndreasPraefcke, Ekki01, Lupo, Parpan05, Paul2, Sidonius

File:Old Swiss Confederation.jpg *Source*: http://en.wikipedia.org/w/index.php?title=File:Old_Swiss_Confederation.jpg *License*: unknown *Contributors*: User:Lupo

File:Acte de Médiation mg 0643.jpg *Source*: http://en.wikipedia.org/w/index.php?title=File:Acte_de_Médiation_mg_0643.jpg *License*: unknown *Contributors*: User:Rama

File:Bern, Federal Palace, 1857.jpg *Source*: http://en.wikipedia.org/w/index.php?title=File:Bern,_Federal_Palace,_1857.jpg *License*: unknown *Contributors*: Ch. Durheim

File:Gotthard Eröffnungszug Bellinzona.jpg *Source*: http://en.wikipedia.org/w/index.php?title=File:Gotthard_Eröffnungszug_Bellinzona.jpg *License*: unknown *Contributors*: User:Parpan05

File:Ulrich Wille.jpg *Source*: http://en.wikipedia.org/w/index.php?title=File:Ulrich_Wille.jpg *License*: unknown *Contributors*: Dbachmann, Kresspahl, Roland zh

File:Bundesrat Conseil Consglio Cussegl.jpg *Source*: http://en.wikipedia.org/w/index.php?title=File:Bundesrat_Conseil_Consglio_Cussegl.jpg *License*: unknown *Contributors*: Swiss Federal Council

File:Switzerland relief location map.jpg *Source*: http://en.wikipedia.org/w/index.php?title=File:Switzerland_relief_location_map.jpg *License*: unknown *Contributors*: User:NordNordWest, User:Sting

Image:Hauteroute.jpg *Source*: http://en.wikipedia.org/w/index.php?title=File:Hauteroute.jpg *License*: unknown *Contributors*: User:Jackph

Image:Sanetsch.jpg *Source*: http://en.wikipedia.org/w/index.php?title=File:Sanetsch.jpg *License*: unknown *Contributors*: User:Terra3

Image:Central Switzerland.jpg *Source*: http://en.wikipedia.org/w/index.php?title=File:Central_Switzerland.jpg *License*: unknown *Contributors*: User:Clare66

Image:Aletschhorn from Konkordiaplatz.jpg *Source*: http://en.wikipedia.org/w/index.php?title=File:Aletschhorn_from_Konkordiaplatz.jpg *License*: unknown *Contributors*: User:Tallin

Image:Joux.jpg *Source*: http://en.wikipedia.org/w/index.php?title=File:Joux.jpg *License*: unknown *Contributors*: User:Earth explorer

Image:Lago di Lugano3.jpg *Source*: http://en.wikipedia.org/w/index.php?title=File:Lago_di_Lugano3.jpg *License*: unknown *Contributors*: User:MadGeographer

File:Bundesrat der Schweiz 2011.jpg *Source*: http://en.wikipedia.org/w/index.php?title=File:Bundesrat_der_Schweiz_2011.jpg *License*: unknown *Contributors*: Swiss Federal Council Photographer: Monika Flückiger

File:Bern 104.jpg *Source*: http://en.wikipedia.org/w/index.php?title=File:Bern_104.jpg *License*: unknown *Contributors*: Pazit Polak

File:Landsgemeinde Glarus 2006.jpg *Source*: http://en.wikipedia.org/w/index.php?title=File:Landsgemeinde_Glarus_2006.jpg *License*: unknown *Contributors*: Adrian Sulc

File:Suisse cantons.svg *Source*: http://en.wikipedia.org/w/index.php?title=File:Suisse_cantons.svg *License*: unknown *Contributors*: User:Pymouss44

File:Wappen Aargau matt.svg *Source*: http://en.wikipedia.org/w/index.php?title=File:Wappen_Aargau_matt.svg *License*: unknown *Contributors*: Sa-se

File:Wappen Nidwalden matt.svg *Source*: http://en.wikipedia.org/w/index.php?title=File:Wappen_Nidwalden_matt.svg *License*: unknown *Contributors*: Sa-se

File:Wappen Appenzell Ausserrhoden matt.svg *Source*: http://en.wikipedia.org/w/index.php?title=File:Wappen_Appenzell_Ausserrhoden_matt.svg *License*: unknown *Contributors*: Sa-se

File:Wappen Obwalden matt.svg *Source*: http://en.wikipedia.org/w/index.php?title=File:Wappen_Obwalden_matt.svg *License*: unknown *Contributors*: Sa-se

File:Wappen Appenzell Innerrhoden matt.svg *Source*: http://en.wikipedia.org/w/index.php?title=File:Wappen_Appenzell_Innerrhoden_matt.svg *License*: unknown *Contributors*: Sa-se

File:Wappen Schaffhausen matt.svg *Source*: http://en.wikipedia.org/w/index.php?title=File:Wappen_Schaffhausen_matt.svg *License*: unknown *Contributors*: (previous version maybe User:Delta-9)

File:Wappen Basel-Landschaft matt.svg *Source*: http://en.wikipedia.org/w/index.php?title=File:Wappen_Basel-Landschaft_matt.svg *License*: unknown *Contributors*: User:WIKImaniac

File:Wappen Schwyz matt.svg *Source*: http://en.wikipedia.org/w/index.php?title=File:Wappen_Schwyz_matt.svg *License*: unknown *Contributors*: -

File:Wappen Basel-Stadt matt.svg *Source*: http://en.wikipedia.org/w/index.php?title=File:Wappen_Basel-Stadt_matt.svg *License*: unknown *Contributors*: Sa-se

File:Wappen Solothurn matt.svg *Source*: http://en.wikipedia.org/w/index.php?title=File:Wappen_Solothurn_matt.svg *License*: unknown *Contributors*: Sa-se

File:Wappen Bern matt.svg *Source*: http://en.wikipedia.org/w/index.php?title=File:Wappen_Bern_matt.svg *License*: unknown *Contributors*: User:Sa-se

File:Wappen St. Gallen matt.svg *Source*: http://en.wikipedia.org/w/index.php?title=File:Wappen_St._Gallen_matt.svg *License*: unknown *Contributors*: Sa-se

File:Wappen Freiburg matt.svg *Source*: http://en.wikipedia.org/w/index.php?title=File:Wappen_Freiburg_matt.svg *License*: unknown *Contributors*: Sa-se

File:Wappen Thurgau matt.svg *Source*: http://en.wikipedia.org/w/index.php?title=File:Wappen_Thurgau_matt.svg *License*: unknown *Contributors*: Sa-se

File:Wappen Genf matt.svg *Source*: http://en.wikipedia.org/w/index.php?title=File:Wappen_Genf_matt.svg *License*: unknown *Contributors*: Sa-se

File:Wappen Tessin matt.svg *Source*: http://en.wikipedia.org/w/index.php?title=File:Wappen_Tessin_matt.svg *License*: unknown *Contributors*: Sa-se

File:Wappen Glarus matt.svg *Source*: http://en.wikipedia.org/w/index.php?title=File:Wappen_Glarus_matt.svg *License*: unknown *Contributors*: Sa-se

File:Wappen Uri matt.svg *Source*: http://en.wikipedia.org/w/index.php?title=File:Wappen_Uri_matt.svg *License*: unknown *Contributors*: Sa-se

File:Wappen Graubünden matt.svg *Source*: http://en.wikipedia.org/w/index.php?title=File:Wappen_Graubünden_matt.svg *License*: unknown *Contributors*: Sa-se

File:Wappen Wallis matt.svg *Source*: http://en.wikipedia.org/w/index.php?title=File:Wappen_Wallis_matt.svg *License*: unknown *Contributors*: Sa-se

File:Wappen Jura matt.svg *Source*: http://en.wikipedia.org/w/index.php?title=File:Wappen_Jura_matt.svg *License*: unknown *Contributors*: Sa-se

File:Wappen Waadt matt.svg *Source*: http://en.wikipedia.org/w/index.php?title=File:Wappen_Waadt_matt.svg *License*: unknown *Contributors*: Sa-se

File:Wappen Luzern matt.svg *Source*: http://en.wikipedia.org/w/index.php?title=File:Wappen_Luzern_matt.svg *License*: unknown *Contributors*: Sa-se

File:Wappen Zug matt.svg *Source*: http://en.wikipedia.org/w/index.php?title=File:Wappen_Zug_matt.svg *License*: unknown *Contributors*: Sa-se

File:Wappen Neuenburg matt.svg *Source*: http://en.wikipedia.org/w/index.php?title=File:Wappen_Neuenburg_matt.svg *License*: unknown *Contributors*: Sa-se

File:Wappen Zürich matt.svg *Source*: http://en.wikipedia.org/w/index.php?title=File:Wappen_Zürich_matt.svg *License*: unknown *Contributors*: Sa-se

File:Geneve2.jpg *Source*: http://en.wikipedia.org/w/index.php?title=File:Geneve2.jpg *License*: unknown *Contributors*: User:MadGeographer

File:Flag of the Red Cross.svg *Source*: http://en.wikipedia.org/w/index.php?title=File:Flag_of_the_Red_Cross.svg *License*: unknown *Contributors*: User:Jhs

File:Axalp FA-18C 5.jpg *Source*: http://en.wikipedia.org/w/index.php?title=File:Axalp_FA-18C_5.jpg *License*: unknown *Contributors*: User:Cooper.ch

File:SKdt-Fahrzeug - Schweizer Armee - Steel Parade 2006.jpg *Source*: http://en.wikipedia.org/w/index.php?title=File:SKdt-Fahrzeug_-_Schweizer_Armee_-_Steel_Parade_2006.jpg *License*: unknown *Contributors*: User:Sandstein

File:Vintage Omega Speedmaster 145.012-67.jpg *Source*: http://en.wikipedia.org/w/index.php?title=File:Vintage_Omega_Speedmaster_145.012-67.jpg *License*: unknown *Contributors*: Shane Lin

File:Switzerland Export Treemap.png *Source*: http://en.wikipedia.org/w/index.php?title=File:Switzerland_Export_Treemap.png *License*: unknown *Contributors*: 1Veertje, MIT Harvard Observatory, 1 anonymous edits

File:2011-06-14 08-28-11 Switzerland Kanton Zürich Zürich Hönggerberg.jpg *Source*: http://en.wikipedia.org/w/index.php?title=File:2011-06-14_08-28-11_Switzerland_Kanton_Zürich_Zürich_Hönggerberg.jpg *License*: unknown *Contributors*: User:Simisa

File:Engadine.jpg *Source*: http://en.wikipedia.org/w/index.php?title=File:Engadine.jpg *License*: unknown *Contributors*: User:Biovit

File:Swiss scientists.jpg *Source*: http://en.wikipedia.org/w/index.php?title=File:Swiss_scientists.jpg *License*: unknown *Contributors*: User:MadGeographer

File:ETHZ.JPG *Source*: http://en.wikipedia.org/w/index.php?title=File:ETHZ.JPG *License*: unknown *Contributors*: User:ZachT

File:LHC, CERN.jpg *Source*: http://en.wikipedia.org/w/index.php?title=File:LHC,_CERN.jpg *License*: unknown *Contributors*: Max Braun on Flickr

File:Akw goesgenmai2010.JPG *Source*: http://en.wikipedia.org/w/index.php?title=File:Akw_goesgenmai2010.JPG *License*: unknown *Contributors*: User:Ch-info.ch

File:Lötschberg Tunnel.jpg *Source*: http://en.wikipedia.org/w/index.php?title=File:Lötschberg_Tunnel.jpg *License*: unknown *Contributors*: User:Ttrainer

File:Sprachen CH 2000 EN.svg *Source*: http://en.wikipedia.org/w/index.php?title=File:Sprachen_CH_2000_EN.svg *License*: unknown *Contributors*: User:Sidonius

File:Bisse de Clavau.jpg *Source*: http://en.wikipedia.org/w/index.php?title=File:Bisse_de_Clavau.jpg *License*: unknown *Contributors*: Londo Mollari

File:Glarus Stadtkirche Langhaus.jpg *Source*: http://en.wikipedia.org/w/index.php?title=File:Glarus_Stadtkirche_Langhaus.jpg *License*: unknown *Contributors*: User:Ikiwaner

File:Vals06.JPG *Source*: http://en.wikipedia.org/w/index.php?title=File:Vals06.JPG *License*: unknown *Contributors*: AndreasPraefcke, Manoillon, Paebi, Wst, 1 anonymous edits

File:Rousseau Geneve.JPG *Source*: http://en.wikipedia.org/w/index.php?title=File:Rousseau_Geneve.JPG *License*: unknown *Contributors*: myself

File:Allalinhorn 4027 2007 04 17.JPG *Source*: http://en.wikipedia.org/w/index.php?title=File:Allalinhorn_4027_2007_04_17.JPG *License*: unknown *Contributors*: ch-info.ch

File:Innenaufnahme Vaillant Arena Davos.JPG *Source*: http://en.wikipedia.org/w/index.php?title=File:Innenaufnahme_Vaillant_Arena_Davos.JPG *License*: unknown *Contributors*: Original uploader was Abercio at de.wikipedia (Original text : Michael Petri)

File:Federer Cincinnati (2007).jpg *Source*: http://en.wikipedia.org/w/index.php?title=File:Federer_Cincinnati_(2007).jpg *License*: unknown *Contributors*: jtbarrett

GNU Free Documentation License Version 1.2, November 2002 Copyright (C) 2000,2001,2002 Free Software Foundation, Inc. 59 Temple Place, Suite 330, Boston, MA 02111-1307 USA Everyone is permitted to copy and distribute verbatim copies of this license document, but changing it is not allowed.

0. PREAMBLE

The purpose of this License is to make a manual, textbook, or other functional and useful document "free" in the sense of freedom: to assure everyone the effective freedom to copy and redistribute it, with or without modifying it, either commercially or noncommercially. Secondarily, this License preserves for the author and publisher a way to get credit for their work, while not being considered responsible for modifications made by others. This License is a kind of "copyleft", which means that derivative works of the document must themselves be free in the same sense. It complements the GNU General Public License, which is a copyleft license designed for free software. We have designed this License in order to use it for manuals for free software, because free software needs free documentation: a free program should come with manuals providing the same freedoms that the software does. But this License is not limited to software manuals; it can be used for any textual work, regardless of subject matter or whether it is published as a printed book. We recommend this License principally for works whose purpose is instruction or reference.

1. APPLICABILITY AND DEFINITIONS

This License applies to any manual or other work, in any medium, that contains a notice placed by the copyright holder saying it can be distributed under the terms of this License. Such a notice grants a world-wide, royalty-free license, unlimited in duration, to use that work under the conditions stated herein. The "Document", below, refers to any such manual or work. Any member of the public is a licensee, and is addressed as "you". You accept the license if you copy, modify or distribute the work in a way requiring permission under copyright law. A "Modified Version" of the Document means any work containing the Document or a portion of it, either copied verbatim, or with modifications and/or translated into another language. A "Secondary Section" is a named appendix or a front-matter section of the Document that deals exclusively with the relationship of the publishers or authors of the Document to the Document's overall subject (or to related matters) and contains nothing that could fall directly within that overall subject. (Thus, if the Document is in part a textbook of mathematics, a Secondary Section may not explain any mathematics.) The relationship could be a matter of historical connection with the subject or with related matters, or of legal, commercial, philosophical, ethical or political position regarding them. The "Invariant Sections" are certain Secondary Sections whose titles are designated, as being those of Invariant Sections, in the notice that says that the Document is released under this License. If a section does not fit the above definition of Secondary then it is not allowed to be designated as Invariant. The Document may contain zero Invariant Sections. If the Document does not identify any Invariant Sections then there are none. The "Cover Texts" are certain short passages of text that are listed, as Front-Cover Texts or Back-Cover Texts, in the notice that says that the Document is released under this License. A Front-Cover Text may be at most 5 words, and a Back-Cover Text may be at most 25 words. A "Transparent" copy of the Document means a machine-readable copy, represented in a format whose specification is available to the general public, that is suitable for revising the document straightforwardly with generic text editors or (for images composed of pixels) generic paint programs or (for drawings) some widely available drawing editor, and that is suitable for input to text formatters or for automatic translation to a variety of formats suitable for input to text formatters. A copy made in an otherwise Transparent file format whose markup, or absence of markup, has been arranged to thwart or discourage subsequent modification by readers is not Transparent. An image format is not Transparent if used for any substantial amount of text. A copy that is not "Transparent" is called "Opaque". Examples of suitable formats for Transparent copies include plain ASCII without markup, Texinfo input format, LaTeX input format, SGML or XML using a publicly available DTD, and standard-conforming simple HTML, PostScript or PDF designed for human modification. Examples of transparent image formats include PNG, XCF and JPG. Opaque formats include proprietary formats that can be read and edited only by proprietary word processors, SGML or XML for which the DTD and/or processing tools are not generally available, and the machine-generated HTML, PostScript or PDF produced by some word processors for output purposes only. The "Title Page" means, for a printed book, the title page itself, plus such following pages as are needed to hold, legibly, the material this License requires to appear in the title page. For works in formats which do not have any title page as such, "Title Page" means the text near the most prominent appearance of the work's title, preceding the beginning of the body of the text. A section "Entitled XYZ" means a named subunit of the Document whose title either is precisely XYZ or contains XYZ in parentheses following text that translates XYZ in another language. (Here XYZ stands for a specific section name mentioned below, such as "Acknowledgements", "Dedications", "Endorsements", or "History".) To "Preserve the Title" of such a section when you modify the Document means that it remains a section "Entitled XYZ" according to this definition. The Document may include Warranty Disclaimers next to the notice which states that this License applies to the Document. These Warranty Disclaimers are considered to be included by reference in this License, but only as regards disclaiming warranties: any other implication that these Warranty Disclaimers may have is void and has no effect on the meaning of this License.

2. VERBATIM COPYING

You may copy and distribute the Document in any medium, either commercially or noncommercially, provided that this License, the copyright notices, and the license notice saying this License applies to the Document are reproduced in all copies, and that you add no other conditions whatsoever to those of this License. You may not use technical measures to obstruct or control the reading or further copying of the copies you make or distribute. However, you may accept compensation in exchange for copies. If you distribute a large enough number of copies you must also follow the conditions in section 3. You may also lend copies, under the same conditions stated above, and you may publicly display copies.

3. COPYING IN QUANTITY

If you publish printed copies (or copies in media that commonly have printed covers) of the Document, numbering more than 100, and the Document's license notice requires Cover Texts, you must enclose the copies in covers that carry, clearly and legibly, all these Cover Texts: Front-Cover Texts on the front cover, and Back-Cover Texts on the back cover. Both covers must also clearly and legibly identify you as the publisher of these copies. The front cover must present the full title with all words of the title equally prominent and visible. You may add other material on the covers in addition. Copying with changes limited to the covers, as long as they preserve the title of the Document and satisfy these conditions, can be treated as verbatim copying in other respects. If the required texts for either cover are too voluminous to fit legibly, you should put the first ones listed (as many as fit reasonably) on the actual cover, and continue the rest onto adjacent pages. If you publish or distribute Opaque copies of the Document numbering more than 100, you must either include a machine-readable Transparent copy along with each Opaque copy, or state in or with each Opaque copy a computer-network location from which the general network-using public has access to download using public-standard network protocols a complete Transparent copy of the Document, free of added material. If you use the latter option, you must take reasonably prudent steps, when you begin distribution of Opaque copies in quantity, to ensure that this Transparent copy will remain thus accessible at the stated location until at least one year after the last time you distribute an Opaque copy (directly or through your agents or retailers) of that edition to the public. It is requested, but not required, that you contact the authors of the Document well before redistributing any large number of copies, to give them a chance to provide you with an updated version of the Document.

4. MODIFICATIONS

You may copy and distribute a Modified Version of the Document under the conditions of sections 2 and 3 above, provided that you release the Modified Version under precisely this License, with the Modified Version filling the role of the Document, thus licensing distribution and modification of the Modified Version to whoever possesses a copy of it. In addition, you must do these things in the Modified Version: A. Use in the Title Page (and on the covers, if any) a title distinct from that of the Document, and from those of previous versions (which should, if there were any, be listed in the History section of the Document). You may use the same title as a previous version if the original publisher of that version gives permission. B. List on the Title Page, as authors, one or more persons or entities responsible for authorship of the modifications in the Modified Version, together with at least five of the principal authors of the Document (all of its principal authors, if it has fewer than five), unless they release you from this requirement. C. State on the Title page the name of the publisher of the Modified Version, as the publisher. D. Preserve all the copyright notices of the Document. E. Add an appropriate copyright notice for your modifications adjacent to the other copyright notices. F. Include, immediately after the copyright notices, a license notice giving the public permission to use the Modified Version under the terms of this License, in the form shown in the Addendum below. G. Preserve in that license notice the full lists of Invariant Sections and required Cover Texts given in the Document's license notice. H. Include an unaltered copy of this License. I. Preserve the section Entitled "History", Preserve its Title, and add to it an item stating at least the title, year, new authors, and publisher of the Modified Version as given on the Title Page. If there is no section Entitled "History" in the Document, create one stating the title, year, authors, and publisher of the Document as given on its Title Page, then add an item describing the Modified Version as stated in the previous sentence. J. Preserve the network location, if any, given in the Document for public access to a Transparent copy of the Document, and likewise the network locations given in the Document for previous versions it was based on. These may be placed in the "History" section. You may omit a network location for a work that was published at least four years before the Document itself, or if the original publisher of the version it refers to gives permission. K. For any section Entitled "Acknowledgements" or "Dedications", Preserve the Title of the section, and preserve in the section all the substance and tone of each of the contributor acknowledgements and/or dedications given therein. L. Preserve all the Invariant Sections of the Document, unaltered in their text and in their titles. Section numbers or the equivalent are not considered part of the section titles. M. Delete any section Entitled "Endorsements". Such a section may not be included in the Modified Version. N. Do not retitle any existing section to be Entitled "Endorsements" or to conflict in title with any Invariant Section. O. Preserve any Warranty Disclaimers. If the Modified Version includes new front-matter sections or appendices that qualify as Secondary Sections and contain no material copied from the Document, you may at your option designate some or all of these sections as invariant. To do this, add their titles to the list of Invariant Sections in the Modified Version's license notice. These titles must be distinct from any other section titles. You may add a section Entitled "Endorsements", provided it contains nothing but endorsements of your Modified Version by various parties--for example, statements of peer review or that the text has been approved by an organization as the authoritative definition of a standard. You may add a passage of up to five words as a Front-Cover Text, and a passage of up to 25 words as a Back-Cover Text, to the end of the list of Cover Texts in the Modified Version. Only one passage of Front-Cover Text and one of Back-Cover Text may be added by (or through arrangements made by) any one entity. If the Document already includes a cover text for the same cover, previously added by you or by arrangement made by the same entity you are acting on behalf of, you may not add another; but you may replace the old one, on explicit permission from the previous publisher that added the old one. The author(s) and publisher(s) of the Document do not by this License give permission to use their names for publicity for or to assert or imply endorsement of any Modified Version.

5. COMBINING DOCUMENTS

You may combine the Document with other documents released under this License, under the terms defined in section 4 above for modified versions, provided that you include in the combination all of the Invariant Sections of all of the original documents, unmodified, and list them all as Invariant Sections of your combined work in its license notice, and that you preserve all their Warranty Disclaimers. The combined work need only contain one copy of this License, and multiple identical Invariant Sections may be replaced with a single copy. If there are multiple Invariant Sections with the same name but different contents, make the title of each such section unique by adding at the end of it, in parentheses, the name of the original author or publisher of that section if known, or else a unique number. Make the same adjustment to the section titles in the list of Invariant Sections in the license notice of the combined work. In the combination, you must combine any sections Entitled "History" in the various original documents, forming one section Entitled "History"; likewise combine any sections Entitled "Acknowledgements", and any sections Entitled "Dedications". You must delete all sections Entitled "Endorsements".

6. COLLECTIONS OF DOCUMENTS

You may make a collection consisting of the Document and other documents released under this License, and replace the individual copies of this License in the various documents with a single copy that is included in the collection, provided that you follow the rules of this License for verbatim copying of each of the documents in all other respects. You may extract a single document from such a collection, and distribute it individually under this License, provided you insert a copy of this License into the extracted document, and follow this License in all other respects regarding verbatim copying of that document.

7. AGGREGATION WITH INDEPENDENT WORKS

A compilation of the Document or its derivatives with other separate and independent documents or works, in or on a volume of a storage or distribution medium, is called an "aggregate" if the copyright resulting from the compilation is not used to limit the legal rights of the compilation's users beyond what the individual works permit. When the Document is included in an aggregate, this License does not apply to the other works in the aggregate which are not themselves derivative works of the Document. If the Cover Text requirement of section 3 is applicable to these copies of the Document, then if the Document is less than one half of the entire aggregate, the Document's Cover Texts may be placed on covers that bracket the Document within the aggregate, or the electronic equivalent of covers if the Document is in electronic form. Otherwise they must appear on printed covers that bracket the whole aggregate.

8. TRANSLATION

Translation is considered a kind of modification, so you may distribute translations of the Document under the terms of section 4. Replacing Invariant Sections with translations requires special permission from their copyright holders, but you may include translations of some or all Invariant Sections in addition to the original versions of these Invariant Sections. You may include a translation of this License, and all the license notices in the Document, and any Warranty Disclaimers, provided that you also include the original English version of this License and the original versions of those notices and disclaimers. In case of a disagreement between the translation and the original version of this License or a notice or disclaimer, the original version will prevail. If a section in the Document is Entitled "Acknowledgements", "Dedications", or "History", the requirement (section 4) to Preserve its Title (section 1) will typically require changing the actual title.

9. TERMINATION

You may not copy, modify, sublicense, or distribute the Document except as expressly provided for under this License. Any other attempt to copy, modify, sublicense or distribute the Document is void, and will automatically terminate your rights under this License. However, parties who have received copies, or rights, from you under this License will not have their licenses terminated so long as such parties remain in full compliance.

10. FUTURE REVISIONS OF THIS LICENSE

The Free Software Foundation may publish new, revised versions of the GNU Free Documentation License from time to time. Such new versions will be similar in spirit to the present version, but may differ in detail to address new problems or concerns. See http://www.gnu.org/copyleft/. Each version of the License is given a distinguishing version number. If the Document specifies that a particular numbered version of this License "or any later version" applies to it, you have the option of following the terms and conditions either of that specified version or of any later version that has been published (not as a draft) by the Free Software Foundation. If the Document does not specify a version number of this License, you may choose any version ever published (not as a draft) by the Free Software Foundation. ADDENDUM: How to use this License for your documents To use this License in a document you have written, include a copy of the License in the document and put the following copyright and license notices just after the title page: Copyright (c) YEAR YOUR NAME. Permission is granted to copy, distribute and/or modify this document under the terms of the GNU Free Documentation License, Version 1.2 or any later version published by the Free Software Foundation; with no Invariant Sections, no Front-Cover Texts, and no Back-Cover Texts. A copy of the license is included in the section entitled "GNU Free Documentation License". If you have Invariant Sections, Front-Cover Texts and Back-Cover Texts, replace the "with...Texts." line with this: with the Invariant Sections being LIST THEIR TITLES, with the Front-Cover Texts being LIST, and with the Back-Cover Texts being LIST. If you have Invariant Sections without Cover Texts, or some other combination of the three, merge those two alternatives to suit the situation. If your document contains nontrivial examples of program code, we recommend releasing these examples in parallel under your choice of free software license, such as the GNU General Public License, to permit their use in free software.

MIX
Papier aus verantwortungsvollen Quellen
Paper from responsible sources
FSC® C105338

Printed by Books on Demand GmbH, Norderstedt / Germany